Vicky Goralczyk

Cells in Ceramics: Finding out About Getting in

AF009592

Vicky Goralczyk

Cells in Ceramics: Finding out About Getting in

Development and Operation of a Perfusion Bioreactor for the Cultivation of Mammalian Cells Inside a Sponge-Like Ceramic Matrix

Südwestdeutscher Verlag für Hochschulschriften

Imprint
Any brand names and product names mentioned in this book are subject to trademark, brand or patent protection and are trademarks or registered trademarks of their respective holders. The use of brand names, product names, common names, trade names, product descriptions etc. even without a particular marking in this work is in no way to be construed to mean that such names may be regarded as unrestricted in respect of trademark and brand protection legislation and could thus be used by anyone.

Publisher:
Südwestdeutscher Verlag für Hochschulschriften
is a trademark of
Dodo Books Indian Ocean Ltd., member of the OmniScriptum S.R.L Publishing group
str. A.Russo 15, of. 61, Chisinau-2068, Republic of Moldova Europe
Printed at: see last page
ISBN: 978-3-8381-2413-1

Zugl. / Approved by: Berlin, Technische Universität Berlin, Dissertation, 2010

Copyright © Vicky Goralczyk
Copyright © 2011 Dodo Books Indian Ocean Ltd., member of the OmniScriptum S.R.L Publishing group

> "Procrastination isn't the problem,
> it's the solution!"
>
> — *Ellen DeGeneres*

Contents

1 **Abstract**   1

2 **Introduction**   3
    2.1 State of the art 4
        2.1.1 Ceramics for cell cultivation 4
        2.1.2 Cell cultivation modes 7
        2.1.3 Inoculation modes 9
        2.1.4 Perfusion dynamic 12
    2.2 Goals of this thesis 12

3 **Materials and methods**   15
    3.1 Preparation of alumina ceramics for cell culture 15
    3.2 Reactor device for cell cultivation on alumina foams 15
        3.2.1 Tubular design for series connection of ceramics 16
        3.2.2 Revolver design for parallel connection of ceramics 18
        3.2.3 Small block design for single foam analysis 19
    3.3 Inoculation procedures 19
        3.3.1 Static inoculation 19
        3.3.2 Dynamic inoculation by stirring or agitation 21
        3.3.3 Dynamic inoculation by convectional forces 21
    3.4 Cultivation of cells on ceramics 21
        3.4.1 Perfusion cultivation inside the reactor device 23
        3.4.2 Static cultivation of cells outside the reactor device 23
        3.4.3 Cultivation of cells by medium convection outside the reactor device . . . 23
    3.5 Cell lines and origin 23
    3.6 Assays 25
        3.6.1 Microscopic evaluation of cell vitality by dyeing according to FDA/EB protocol 25
        3.6.2 Microscopic evaluation of cell distribution by dyeing according to hematoxylin/eosin protocol 26
        3.6.3 Scanning electron microscopy of ceramic surface and cells on ceramics . . 26
        3.6.4 Metabolic evaluation of glucose consumption and lactate formation . . . 26
        3.6.5 Reduction of resazurin 27
        3.6.6 Carrier hot gas extraction 29

4 Reactor characterization · 31
4.1 Ceramics · 31
4.1.1 Porosity · 31
4.1.2 Flow resistance · 31
4.2 Characterization of flow inside the reactor · 33

5 Analysis of *CHO-K1* by resazurin assay and carbon content determination. · 37
5.1 Reduction of resazurin by *CHO-K1* · 37
5.1.1 Determination of rate of reduction · 37
5.1.2 Modeling resazurin reduction · 40
5.1.3 Adaptation of the resazurin reduction model to describe bioreactor performance · 41
5.2 Carbon content of cells cultivated on ceramics · 46
5.2.1 Carbon content of pure cells · 46
5.2.2 Carbon content of cells on foams · 47

6 Influence of mode of inoculation on cellular growth and distribution · 51
6.1 Static inoculation · 51
6.1.1 Static inoculation on foams in culture plates · 51
6.1.2 Static inoculation into the tubular reactor · 52
6.1.3 Static inoculation into revolver reactors · 53
6.1.4 Static inoculation in ceramics with flow channels · 55
6.1.5 Reproducible cultivation following static inoculation into revolver reactors · 60
6.2 Dynamic inoculation by agitation · 62
6.3 Oscillatory perfusion inoculation · 65
6.3.1 Influence of module orientation, flow velocity, initial cell count on cell distribution · 65
6.3.2 Introduction of more porous ceramics and augmentation of flow velocity during cultivation · 66
6.3.3 Reduction of foam volume by reducing cylinder height · 70

7 Reproducibility of the chosen operation methods · 73
7.1 Standard foams · 73
7.2 Foams with larger pores · 75

8 Long-term cultivation · 83

9 Applicability of the reactor system for other cell types · 89
9.1 Human lung carcinoma cells *A549* · 89
9.2 Human primary fibroblasts · 91
9.3 Madin-Darby canine kidney cells (*MDCK*) · 91

10 Conclusion · 97

Appendix 101

Literature 114

Abbreviations

Al_2O_3 aluminium oxide

CO_2 carbon dioxide

CHGE carrier hot gas extraction

DNA desoxyribonucleic acid

EB ethidium bromide

ECM extracellular matrix

FBS fetal bovine serum

FDA fluorescein diacetate

HA hydroxy-apatite

IFS Interdisciplinary Research Priority Program

PBS++ phosphate buffered saline containing magnesium and calcium

PFR plug flow reactor

ppi pores per inch

SD standard deviation

SEM scanning electron microscopy

STR stirred tank reactor

TCP tri-calcium phosphate

TU Berlin Technische Universität Berlin

Symbols

Δp	pressure drop [Pa]
$\dot{V}$	volumetric flow [ml/min]
η	coefficient of viscosity [mPas]
ϕ	porosity
ρ	density [g/ml]
σ	standard deviation
θ	normalized time
A	area [m^2]
Bo	Bodenstein number
c	concentration [g/l]
D_{ax}	axial coefficient of dispersion [m^2/s]
E	extinction
F	normalized extinction
H	height [m]
K_d	Darcy's constant [m^2]
L	length [m]
m	mass [g]
n	quantity
p	percentage
R^2	coefficient of determination
t	time [s]
V	volume [ml]
x	normalized concentration

List of Figures

2.1	Cellular embedment in connective tissue.	5
2.2	Al_2O_3 structure and surface.	6
2.3	One foam replaces 6 T-flasks.	7
3.1	Pore size distribution of foamed alumina.	16
3.2	Module connection configurations.	17
3.3	Tubular design reactor module for series connection of foams.	18
3.4	Tubular configuration connected to the Biostat B plus.	18
3.5	Ceramic foam holding magazines.	19
3.6	Reactor modules for parallel perfusion.	20
3.7	Single foam reaction module.	20
3.8	Reactor module with glass olives.	22
4.1	Experimental setup for evaluation of K_d.	32
4.2	Color step experiment.	35
5.1	Dynamic of resazurin reduction for *CHO-K1*.	38
5.2	Relation between rate of reduction and cell number.	40
5.3	Model fit for reduction of resazurin for identification experiments.	42
5.4	Model validation for reduction of resazurin for non-identification experiments.	43
5.5	Scheme for division of reactor volume into compartments for flow modeling.	44
5.6	Model validation for reduction of resazurin inside the bioreactor.	47
5.7	Carbon and sulfur content of *CHO-K1*.	47
5.8	Carbon content of *CHO-K1* on ceramics.	48
6.1	Resazurin reduction in static approach.	52
6.2	Cell leakage for cells statically inoculated and cultivated with agitation.	52
6.3	Resazurin reduction after static inoculation in reactor module.	53
6.4	Static inoculation of cells into revolver reactor.	54
6.5	Medium metabolites for static inoculation into reactor moules.	55
6.6	FDA/EB staining for static inoculation in reactor modules.	56
6.7	Medium metabolites for foams with blind holes.	57
6.8	Medium single foam analysis for blind hole foams.	58
6.9	Resazurin single foam assay for blind hole foams.	58
6.10	Vitality assay for blind hole foams.	59

6.11	Dynamic resazurin reduction after static inoculation in reactor module.	60
6.12	Medium metabolites for static inoculation into conical magazines.	61
6.13	Scanning electron microscopy of *CHO-K1* on ceramic foam.	61
6.14	Cell count for agitation inoculation.	62
6.15	Cell density and distribution following agitation seeding.	64
6.16	Cell distribution following dynamic seeding by help of perfusion.	66
6.17	Scanning electron microscopy following dynamic seeding by help of perfusion.	67
6.18	Comparison of resazurin reduction for small and wider pore sizes.	68
6.19	Glucose consumption and lactate formation for different pore sizes.	69
6.20	Reduction of resazurin for flat foams with larger pores.	71
6.21	Vitality assay for *CHO-K1* on flat ceramic foams with larger pores.	71
7.1	Cumulative glucose consumption and lactate formation for reproducible inoculation of *CHO-K1* on standard foams.	74
7.2	Reduction of resazurin for *CHO-K1* reproducibly inoculated on standard foams.	75
7.3	Cell distribution for *CHO-K1* reproducibly inoculated.	76
7.4	Cumulative glucose consumption and lactate formation for reproducible inoculation of *CHO-K1* on flat foams with larger pores.	77
7.5	Reduction of resazurin for *CHO-K1* reproducibly inoculated on flat foams with larger pores.	78
7.6	Cell distribution for *CHO-K1* reproducibly inoculated on flat foams with larger pores.	79
7.7	Scanning electron microscopy for *CHO-K1* on flat foams with larger pores.	80
8.1	Cumulative glucose consumption and lactate formation for long-term cultivation of *CHO-K1*.	84
8.2	Metabolic reduction of resazurin by cells cultivated on standard foams for 7 weeks under medium perfusion.	85
8.3	Cell proliferation for long-term cultivation.	85
8.4	Vitality staining for *CHO-K1* cultivated for several weeks in standard foams and foams with larger pores.	86
8.5	Scanning electron microscopy for long-term cultivation of *CHO-K1*.	87
9.1	Vitality staining of *A549* following 2 weeks of cultivation.	90
9.2	Scanning electron microscopy of *A549* following 2 weeks of cultivation.	91
9.3	Vitality staining of primary fibroblasts following 7 days of cultivation.	92
9.4	Scanning electron microscopy of primary fibroblasts following 7 days of cultivation.	93
9.5	Staining of *MDCK* following 2 weeks of cultivation.	94
9.6	Scanning electron microscopy of *MDCK* following 2 weeks of cultivation.	95
10.1	Illustration of self-made shaker.	102

List of Tables

2.1 Requirements of scaffolds used for cell cultivation in packed bed bioreactors [Meuwly et al., 2007]. 5

3.1 Demands for bioreactor design. 17
3.2 Cultivation requirements for *CHO-K1*, *A549*, *MDCK* and human fibroblasts. . . 24

4.1 Flow distribution inside ceramics. 34
4.2 Characteristic numbers for revolver reactor design. 35

5.1 Linear regression for reduction of resazurin. 39
5.2 Conditions for modeling resazurin reduction inside bioreactors. 45
5.3 Linear regression for carbon analysis. 49

6.1 Agitation inoculation for different cell numbers. 63
6.2 Optical analysis of foams two days after perfusion inoculation. 65
6.3 Optical analysis of flat foams two days after perfusion inoculation. 70

7.1 Reproducibility of cell cultivation in reactor modules. 76

9.1 Optical analysis of *A549* on ceramics three days after perfusion inoculation. . . 90
9.2 Optical analysis of primary fibroblasts on ceramics 7 days after perfusion inoculation. 91
9.3 Optical analysis of *MDCK* on ceramics three days after perfusion inoculation. . 94

10.1 Adaptation of *CHO-K1* to growth in medium containing 1% FBS. 105

1 Abstract

Cultivation of adherently growing cells inside three dimensional scaffolds gains more importance as specific demands from medicine and industry arise. A huge challenge herein lies in nutrient supply all over the scaffold's volume. This work presents a perfusion bioreactor device for cell cultivation inside porous aluminium oxide ceramics. The bioreactor supports vital growth of *CHO-K1*, *A549*, *MDCK* and human primary fibroblasts on and in ceramic cylinders of 5 mm height and 10 mm diameter with perfusion velocities from 0.33-0.66 ml min^{-1} cm^{-2}. Proliferation was assured by a metabolic assay based on resazurin reduction and carbon content analysis. Cell distribution and vitality was assessed by staining cross sections of the ceramic.

Fibrous structures were observed inside ceramic scaffolds for perfusion cultivation of *CHO-K1* and human primary fibroblasts for several weeks by evaluation of scanning electron microscopy pictures. This clearly supports the hypothesis of micro milieu formation inside porous ceramics and consequently abundance of ECM-proteins.

Perfusion therefore enables three-dimensional cell cultivation inside porous alumina ceramics.

2 Introduction

In the 1980s, ceramics were discovered as scaffolds for cell cultivation [Lydersen et al., 1985] for the first time. Lydersen in his frequently cited paper proved the feasibility of the scaffold to support growth of different cell types as mammalian cells, insect cells, chicken and primate cells. In the following years, other groups also used scaffolds with a plain ceramic surface for cell cultivation [Lydersen et al., 1985, Mitsuda et al., 1991, Suck et al., 2008]. In contrast, a ceramic with a highly interconnected porous structure not only supports cellular growth but moreover was found to promote cellular differentiation [Schubert et al., 2004]. It is assumed, that the porous structure allows the creation of a micro environment favorable for cell differentiation, as cells presumably are exposed to various growth conditions depending on surrounding nutrient status. Since then, porous ceramics have been used as cultivation surface for proliferation and differentiation of production cell lines [Park and Stephanopoulos, 1993], many kinds of primary cells [Wang et al., 2003, Schubert et al., 2004, Janssen et al., 2006] and stem cells [Wang et al., 2003, Janssen et al., 2006].
The great potential of interconnected structured ceramics is in its ability to provide a number of variable micro milieus in different pores at once. Therefore, the formation of a well defined extracellular matrix is stimulated which is an inevitable step in order to cultivate cells under natural conditions [Minuth et al., 2003]. Moreover, the huge surface to volume ratio of a porous structure spares the need for permanent cell passaging. Extensive cell passaging is a standard procedure in cell cultivation techniques but it leads to cellular dedifferentiation [Bonassar and Vacanti, 1998], which should be avoided in many applications. Promotion of a natural extracellular matrix together with the option to grow a huge number of cells without phenotypical changes makes porous ceramics useful for tissue engineering applications or for production of specific drugs or therapeutics. As demands for very specific therapeutics and tissue engineered constructs are rising, simple ceramics turn out to be a great alternative to other systems. The ceramics used in this work are unique in such a manner that the manufacturing process differs strongly from that of porous ceramics used to date. For the commonly used "Schwarzwalder process", forming of ceramics is realized by coating a porous base body in ceramic slurry and removing the base body by heat, which leaves a double-layered structure of pores and bridges. Instead, the unique technique developed in the group of H. Schubert, Technische Universität (TU) Berlin, see [Garrn et al., 2004], uses a ceramic-protein-slurry for foam forming, which leaves a more solid porous ceramic due to the one-layered structure obtained by the process.
The purpose of the work in hand is now to answer the question of maintaining a useful quantity of cells in a vital, proliferating status deep inside a ceramic scaffold obtained by

the novel process described above. Of particular interest would be a three-dimensional (3D) system which provides high ingrowth into the scaffold due to nutrient perfusion. At the same time diffusion-controlled mass transport perpendicular to the main flow direction into the caverns should ensure the setup of local micro milieus. For cells statically cultivated on ceramics (which is done mainly to study cells under close to physiological conditions) [Orlandi et al., 1997, Bagley et al., 1999, Schubert et al., 2004], cellular ingrowth is limited to a few hundred microns. Therefore, to lure cells into scaffolds, these can be overflown with nutrient solution hereby increasing nutrient exchange rate [Wang et al., 2003] or perfused with nutrients which enhances cellular ingrowth strongly. Often, perfusion approaches have to deal with severe inhomogeneities in cellular distribution and therefore inconsistent cultivation outcomes [Mitsuda et al., 1991, Park and Stephanopoulos, 1993, Janssen et al., 2006], which is ascribed to poorly adjusted operational strategies in many cases. Moreover, upscaling of those perfused systems cannot be achieved by simply enlarging the ceramic volume as due to the scaffold's structure perpetuation of specific micro milieus will be impeded. Therefore, for many systems the quantity of cells growing into a specific, environmentally defined compartment is strongly limited as will be discussed below.

2.1 State of the art

2.1.1 Ceramics for cell cultivation

The human body consists of more than 200 cell types, which differ in form and localization depending on their function in the organism [Alberts et al., 2007]. Among those, less than 20 cell types are found to be free-moving through the organism, i.e. blood cells and cells of the immune system. The vast majority of cells is organized in a three-dimensional environment - the extracellular matrix - where the cells communicate with each other and the surrounding environment by chemical signals (see figure 2.1)[1].
This embedded cell organization, which is mainly imprinted by the structure of extracellular proteins, proteoglycans and polysaccharides, in turn models the cell's shape and hereby strongly affects the intracellular spatial organization of cytoplasmic components. Therefore, intracellular structural proteins, functional units and cytoplasmic sugars build up a highly organized compartment itself whose geometric constitution impacts cellular processes.
Sustaining this cytoplasmic order during *in vitro* cultivations of cells demands sophisticated cultivation procedures as well as adequate scaffolds for cell attachment. Meuwly and colleagues in [Meuwly et al., 2007] recently proposed a list of material properties relevant for cell cultivation, which are summarized in table 2.1. Recent advances for cell cultivation scaffolds allowing three-dimensional cellular growth mainly are focused on polymers as poly-lactic acid (PLA) or poly-glycolic acid (PGA). These are rather soft but pressure-resistent materials possessing good manufacturing properties and therefore facilitate the manufacturing of scaffolds with various shapes and internal structures [Vunjak-Novakovic et al., 1996, Marler et al., 1998,

---
[1]For color images visit http://www.ub.tu-berlin.de and search for the online version of this thesis.

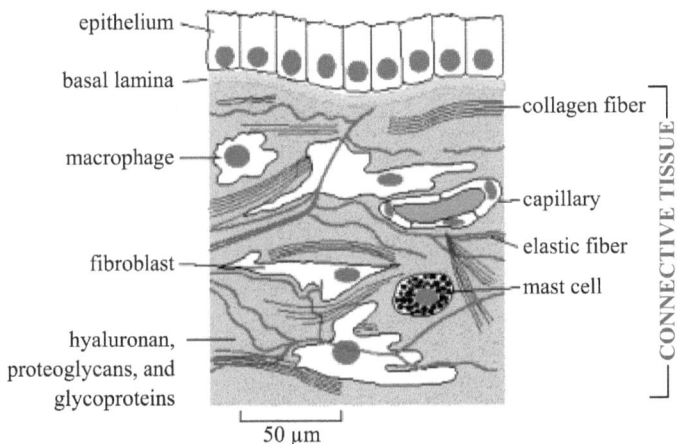

Figure 2.1: Cellular embedment in connective tissue. Adapted from [Alberts et al., 2007].

Table 2.1: Requirements of scaffolds used for cell cultivation in packed bed bioreactors [Meuwly et al., 2007].

simple physical configuration and made of non-toxic materials
high surface to volume ratio
optimal diffusion from the bulk phase to the center of the carrier
chemical and mechanical stability
autoclavable
suitable for adherent and non-adherent cells
chemically and biologically inert, no reaction with the product
low cost and reusable if possible
of nonanimal origin

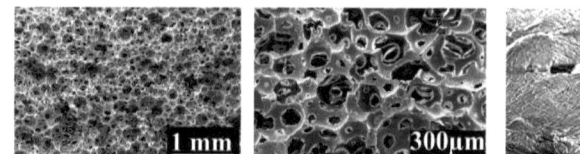

Figure 2.2: Scanning electron micrograph of foamed alumina's structure and surface. Take note of the plaster stone-like surface structure in the last image which is owed to the manufacturing process where small Al_2O_3 particles are pressed against each other.

Kitagawa et al., 2006, Moutos et al., 2007]. Further reported growth substrates are further polymers, hydrogels or hybrid materials as mineralized collagen [Vunjak-Novakovic et al., 1996, Leach and Schmidt, 2004, Liao and Cui, 2004, Moutos et al., 2007].

Opposite to those organic materials, fewer approaches concentrate on anorganic solid bodies which often display a more homogeneous structure and allow an easier and more controllable fabrication process. Herein, the group of ceramic substrates, in particular hydroxy-apatite ceramics (HA) or tri-calcium phosphate (TCP), is most commonly known, as it is used in clinical applications as dental prosthesis or implantation devices mostly for bone repair. Both HA and TCP are biodegradable, meaning the stiff ceramic material is little-by-little replaced by tissue and organism cells of the acceptor body [Krajewski et al., 1996, LeGeros, 2002]. Meanwhile, the anorganic material supports cell proliferation and differentiation, therefore encouraging physical regeneration.

Unfortunately, the decomposition properties of HA and TCP coincidently expose their greatest challenge: depending on porosity and web thickness, ceramics can be very fragile which affects their usability. Moreover, calcium particles disengaging from the ceramic are known to evoke inflammation *in vivo* which of course has to be avoided not only in immuno deficient patients [Harada et al., 1996].

As an alternative to calcium ceramics, aluminium oxide (Al_2O_3) ceramics are reported as cell cultivation matrices since the nineteen-eighties, when Lydersen proved the feasibility to cultivate various cell types on and in the non-modified porous scaffold [Lydersen, 1987]. Due to the fabrication process, foamed alumina represents a sponge-like three-dimensional matrix with huge porosity of around 90%. Moreover, number and diameter of pore openings can be adjusted in a way, that cells in the foam's caverns can be supplied with nutrients by perfusion of medium through the scaffold, see figure 2.2 for an overview of the foam's structure (for ceramics used in this work).

With a surface area to volume ratio of around $0.3\,m^2/ml$ foamed alumina provides a growth area for many cells in just a small reaction volume. One ceramic foam cylinder of 5 mm height and 10 mm diameter could replace 6 in laboratories commonly used $175\,cm^2$-T-flasks regarding cultivation surface, see figure 2.3.

The alumina foam's characteristic purity with respect to organic material is ensured by the last step of the fabrication process. During the sintering process temperatures of around 1650°C are set which incinerate any impurities that may have arisen during former fabrica-

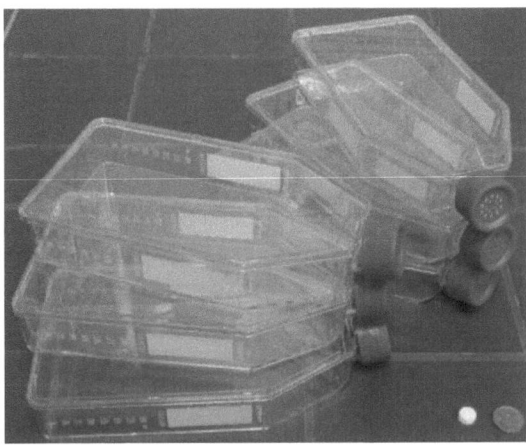

Figure 2.3: Comparison of cultivation surface: One alumina foam cylinder of 5 mm height and 10 mm diameter as used in this work (foam situated left to the € 0.10 coin) can replace 6 175 cm^2 cell culture flasks.

tion [Garrn et al., 2004].
Recent publications clearly indicate that the sum of the alumina foam's beneficial properties promotes cell differentiation in the foam's caverns - without further modification of the ceramic surface or medium manipulation [Schubert et al., 2004].

The presented qualification of aluminium oxide ceramic foams for the cultivation of mammalian cells as well as its appropriateness for cultivations in perfusion mode establishes a basis for the experiments in this thesis. I propose the hypothesis that cellular growth is not only easily achieved on the surface of porous ceramics but also can be maintained *inside* a ceramic foam by help of perfusion of nutrients through the scaffold for long-term cultivation.

2.1.2 Cell cultivation modes

Animal cells *in vivo* are embedded in a pattern composed of polysaccharides and fibrous proteins, the extracellular matrix (ECM). Here, cells are supplied with nutrients and signal substances via a circulating blood system, which is separated from the ECM by a layer of epithelial cells (endothel) and basal lamina, see figure 2.1.
Thus, it appears that nutrients primarily approach convectively with the blood flow, however, secondarily nutrients diffuse through the ECM to the consumpting cells. Physiological studies in vertebrates revealed a maximal diffusion distance between ECM-embedded cells and capillaries of 50-100 µm and with capillary diameters of around 6 µm [Alberts et al., 2007].

For *in vitro* cultivations, the challenge of nutrient supply of cells in monolithic columns of three-dimensional substrates is addressed by a set of different operation strategies as follows.

Static nutrient supply

The most simple operation strategy is to supply cells growing in a three-dimensional scaffold by simply holding the scaffold in a non-moved nutrient solution (static approach). Hereby, nutrients reach the cells solely by diffusion which is driven by the nutrient gradient between near-cell-environment and the solution and which is maintained as long as the gradient upholds. Due to oxygen diffusion limitations, cell growth inside the scaffold is limited to a few hundred microns [Fassnacht and Poertner, 1999, Martin and Vermette, 2005].

In a static environment, equalization of concentration profiles is a rather slow event as it is restricted by diffusion rates.

Nutrient supply by convection

While static operation is acceptable for cell cultures in monolayer, it clearly cannot support cellular growth in three-dimensional structures. To overcome the lack of cells growing in the scaffold's interior, dynamic operation strategies were proposed, in which scaffolds are flooded with medium in a way that increases the probability of medium accessing the scaffold's inward. Hereby, two different approaches can be distinguished: Either medium is conducted over a firmly anchored scaffold [Suzuki et al., 1994, Vunjak-Novakovic et al., 1996] continuously, or the scaffold is moved through the more or less unruffled nutrient solution [Vunjak-Novakovic et al., 1998, Neves et al., 2005, Timmins et al., 2007]. In a mixed approach, scaffolds are exposed to stationary free fall in a rotating, so called microgravitational reactor, favoring the penetration of medium at the scaffold's skin face [Freed and Vunjak-Novakovic, 1997]. To increase shear forces that augment the process of convection, the scaffold can be moved through a phase boundary as with Zellwerk's Z® RP system (Zellwerk GmbH, Eichstädt, Germany). Here cell populated ceramics are moved consecutively through a liquid medium phase and a gaseous air phase, hereby increasing oxygen nourishment compared to submerged cultivations.

Penetration of medium into the flooded scaffold will apparently depend on scaffold porosity (i.e. pore size) and enforcement. The inner volume of a scaffold with small pores therefore requires intense fluid forces, which can be realized in highly oscillating systems.

Compared to static cultivation methods, spatial nutrient concentration will be rather homogeneous, due to well mixing. Local gradients in nutrient composition due to cell metabolism therefore do not establish near cells but will be equalized over the total medium rapidly.

Nutrient supply by perfusion

In order to reach cells which grow deep inside the scaffold's volume, the scaffold has to be perfused with medium. Such an enforced convection, however, requires the sterile incorporation of the scaffold into an operation periphery of tubings and pumps. Then, two principle operational strategies can be distinguished: *i)* medium is pumped through the scaffold into a medium reservoir and thereby recirculated over the volume (recirculation) or *ii)* after one pass over the scaffold volume medium is discarded (single pass), allowing a more or less homogenous nutrient distribution over the cultivation period. The first strategy is more common for economical reasons, and concurrently supports the establishment of a cell-defined signal substance environment as essential cell products are retained in the system.

Applegate and Stephanopoulos in [Applegate and Stephanopoulos, 1992] proved for a perfused 10 cm ceramic monolith that medium recirculation led to enhanced cell death inside the system. Cell mortality mainly was deduced to oxygen limitations which are associated to bed height. Nevertheless, total cell number still was increased compared to single-pass mode [Grampp et al., 1996].

In perfused packed-bed bioreactors, the growing demand for nutrients, which is related to cellular growth, is considered by augmenting medium flow velocity. Therefore, mass throughput is augmented, but likewise laminar or even turbulent shear stresses increase, establishing very new environments for the cells. To isolate cells from the nutrient flow therefore might be a good solution and is addressed in hollow-fiber bioreactors which are known since first experiments by Knazek et al. in the 1970s [Knazek et al., 1972]. Hereby nutrients flow through artificial capillaries and reach cells which grow on the outside of the capillaries by diffusion. This approach resembles more the nutrient supply in organisms but has to deal with high metabolite gradients and accumulation of by-products in the cell-space [Piret et al., 1991] due to large diffusion distances or inhomogenous flow through the hollow fibers which leads to inhomogenous cell growth and often necrosis [Knazek et al., 1972, Ku et al., 1981, Tharakan and Chau, 1986].

Both perfused packed-bed reactors and hollow fiber reactors have to overcome up-scaling difficulties. The height of the bed and the length of the system respectively are strongly limited by axial concentration gradients for nutrients and oxygen. Moreover, the bed diameter for packed-bed bioreactors is limited due to dispersion effects, which influence flow homogeneity.

Nevertheless, bioreactors for cell cultivation using direct perfusion in general promote growth and proliferation inside cell containing scaffolds [Goldstein et al., 2001, Navarro et al., 2001, Bancroft et al., 2002]. Moreover, they have shown very promising cultivation results for many kinds of cells including differentiated primary cells which expressed tissue-specific markers *in vitro* [Pazzano et al., 2000, Goldstein et al., 2001, Bancroft et al., 2002, Carrier et al., 2002, Davisson et al., 2002]. However, perfusion velocity has to be carefully adjusted depending on cultivated cell type and maturation stage [Davisson et al., 2002].

2.1.3 Inoculation modes

Recent work revealed a strong relationship between a scaffold's initial cellularity (i.e. cell number and distribution directly after seeding) and distribution of tissue within an engineered scaffold following further cultivation [Kim et al., 1997, Holy et al., 2000]. Therefore, cell inoculation into three dimensional scaffolds mainly pursues two goals to enhance the rate of tissue development: spatially homogeneous distribution of cells all over the scaffold's volume and an initial cell density high enough to support excellent growth during further cultivation. Hereby inoculation efficieny defined as

$$\text{inoculation efficiency} = \frac{\text{cell number in scaffold following inoculation}}{\text{inoculated cell number}} \quad (2.1)$$

or

$$\text{inoculation efficiency} = 1 - \text{ratio of non attached to attached cells} \quad (2.2)$$

is a major quality factor as cell availability is often limited (i. e. when expanding primary human cells) [Radisic et al., 2003]. The choice of a suitable cell inoculation method is a crucial step for cell cultivation whereby aggressive inoculation methods should be avoided in particular for three dimensional scaffolds as occurring shear stresses can reduce cell viability critically. Therefore, one can distinguish three main strategies, which will be faced in more detail: *i)* static inoculation, where cells are inoculated on the scaffold's surface and allowed to settle before cultivation starts, *ii)* dynamic inoculation facilitated by stirring or shaking the scaffold in a cell suspension, *iii)* dynamic inoculation promoted by convective flow of the cell suspension through the scaffold.

Static inoculation

For static inoculation, scaffolds are either deposited in a cell suspension [Wang et al., 2003] or the cell suspension is dropped on top of the scaffold [Dar et al., 2002]. Cell migration into the porous structure is slightly amplified by capillary action, particularly when scaffolds were thoroughly dried before inoculation [Dar et al., 2002]. Cell leakage is rather unlikely as cells are entrapped by the scaffold's tortuous structure.

Static inoculation methods often are associated with low seeding efficiency [Kim et al., 1997, Xiao et al., 1999, Holy et al., 2000, Li et al., 2001, Kitagawa et al., 2006]. Moreover, they lead to an irregular cellular distribution over the scaffold's volume [Vunjak-Novakovic et al., 1996, Kim et al., 1997, Holy et al., 2000, Li et al., 2001, Wendt et al., 2003]. Dense cell layers are reported on top and bottom sides of the scaffold [Du et al., 2008] with maximal ingrowth depth at about 1 mm [Wendt et al., 2006]. Also, cell aggregation is observed [Dar et al., 2002], which is an important issue as cell spheroids often contain necrotic centers [Sutherland et al., 1986]. Cell constructs maturing from statically inoculated scaffolds are often reported to have an inhomogeneous structure which consists of dense cell layers and concentrated ECM along the periphery together with necrotic interior regions [Wendt et al., 2006]. Static inoculation therefore should be avoided unless for cell types that are very sensitive to mechanical forces [Xiao et al., 1999].

Dynamic inoculation by stirring or agitation

Dynamic inoculation without perfusion of the scaffold often is performed in stirred flask bioreactors (spinner flasks) [Kim et al., 1997, Carrier et al., 1999, Wendt et al., 2003], whereby either both cells and scaffolds are combined in a stirred solution or the constructs are mounted inside the reactor and the cell suspension is stirred around it. Another option is to combine cells and scaffolds in a cultivation tube and to shake it on an agitator [Kim et al., 1997]. In addition to

the physical effects described above, cell migration into the scaffold is supported by convective flow occuring at the cell suspension/scaffold surface boundary.
Mixing inoculation is related to low seeding efficiency [Kim et al., 1997, Carrier et al., 1999] and non-uniform cell distribution [Wendt et al., 2003] at the most. However, [Ouyang and Yang, 2007] recently observed high seeding efficiencies with first order kinetics for spinner flasks. For both mixing and agitation seeding, cells are found inside the scaffold with much higher densities of cells lining the scaffold's surface [Vunjak-Novakovic et al., 1998]. Moreover, small cell aggregates and cell clusters have to be remarked [Kim et al., 1997, Vunjak-Novakovic et al., 1998]. As stated by [Kim et al., 1997], agitation seeding results in higher seeding efficiencies than mixing seeding.
Directly compared to static inoculation methods, mixing and agitation seeding result in a higher adherent cell number inside the scaffold and better cell uniformity [Kim et al., 1997] and are proven to promote a more tissue-like behavior, e. g., for engineered cartilage constructs [Vunjak-Novakovic et al., 1996].

Dynamic inoculation by convectional forces

For both static and mixing/agitation assisted seeding, following inoculation, the cell containing scaffold has to be assembled in a cultivation device when perfusion cultivation is pursued. This critical and potentially non-sterile step is avoided in most dynamic seeding approaches utilizing perfusion, as inoculation is carried out in a combined bioreactor setup which allows cultivation directly after inoculation without any change of the setup [Sodian et al., 2002, Radisic et al., 2003, Wendt et al., 2003, Kitagawa et al., 2006]. For dynamic inoculation by convectional forces the scaffold is perfused with the cell suspension, whereby flow direction can be alternated [Radisic et al., 2003, Wendt et al., 2003], oxygen supply might be regulated [Wendt et al., 2006] and even horizontal rotation of the reactor device is possible [Janssen et al., 2006]. Cell migration into the scaffold's interior, as it is coupled to flow through the scaffold therefore is strongly promoted with cell scaffold contacts being more likely than in static approaches [Xiao et al., 1999]. Apparently, this dynamic approach also causes higher mechanical forces than non-perfusing methods which may be intolerably high for some cell types [Xiao et al., 1999] but can be beneficial for others [Gooch et al., 2001].
If the scaffold's pore size is very small and initial seeding density rather high, one often speaks of filtration seeding, as cells are entrapped in the scaffold's structure as in a filtration unit [Li et al., 2001].
Compared to the seeding methods above, following dynamic inoculation cell distribution is more uniformly [Li et al., 2001, Wendt et al., 2006] with higher seeding efficiencies both for alternating directions [Wendt et al., 2003] and unidirectional inoculation [Zhao and Ma, 2005]. Additional horizontal rotation is reported to yield even more promising results, yet depending on the inner scaffold geometry [Janssen et al., 2006]. Filtration seeding leads to very high cell numbers which seems to hinder cells from further proliferation (a phenomenon known as community effect) [Li et al., 2001]. All in all perfused inoculation results in a good distribution of

cells all over the scaffold with high cell vitality [Du et al., 2008], but can also yield a gradient in cell number from top to bottom [Janssen et al., 2006].

When applying dynamic perfusion inoculation, flow velocity carefully has to be adjusted: high flow rates on the one hand wash dead and unattached cells out of the scaffold [Radisic et al., 2003]. If the flow is too high, on the other hand, already entrapped cells might also be washed out [Kitagawa et al., 2006]. For alternating flow directions flow velocity must not be too low to cope with gravitational forces as otherwise cells will settle at the bottom of the inoculation unit [Wendt et al., 2003, Timmins et al., 2007].

Besides initial cell number which has to be chosen wisely as seeding efficiency drops with higher cell numbers [Li et al., 2001], initial cell density (i. e. cell number divided by volume utilized for perfusion) is an important parameter, hereby it is suggested to use as less volume as possible [Kim et al., 1997, Du et al., 2008].

2.1.4 Perfusion dynamic

For both dynamic perfusion assisted cell inoculation and perfusion cultivation adjusting the medium flow profile is crucial. Cell supply with nutrients and oxygen as well as removal of waste substances has to be maintained at a specific level in order to support cellular growth best. In a porous perfused system, exchanging processes are ascribed to a superposition of convection and diffusion. The rate of exchange can be easily manipulated by applying laminar or turbulent flow and naturally by tuning flow velocity. [Kitagawa et al., 2006] suggest to adjust flow velocity for inoculation and cultivation independently, using a slower profile during inoculation to decrease cell wash out and a higher flow for cultivation, where sufficient nutrient supply is the most crucial issue. Clearly, there is an optimum range for flow velocity beyond which cells will either die because of lack of nutrients or due to intolerable shear forces.

As nutrient demand will rise during cellular growth, it should be considered to control flow velocity during cultivation. This is particularly important when cultivating in single-pass mode as economical considerations dictate flow velocity to be as low as possible.

2.2 Goals of this thesis

Regarding findings as discussed above, among others, the following questions are addressed: How can we unite cells, ceramic scaffolds and a reaction device in a way that cells are grown in a large scaffold volume under steady conditions? Which are the main influencing factors for cellular growth and distribution in a porous ceramic and how do they have to be adjusted to provide cells with a micro milieu they can grow in as close to natural conditions as possible? To answer these questions, scientists of multiple disciplines have joined together in IFS3/2, a TU Berlin funded project. Members were responsible for designing a reactor device (Department of Design Methodology), composing the scaffold's structure (Department for Food Chemistry) and fabricating a porous ceramic (Department of Ceramics) that satisfies all demands for cell cultivation (performed at Chair of Measurement and Control).

2. Introduction

The current work presents part of the progress of the ambitious project followed up by the last groupThe herein pursued main goal of cultivating cells inside the porous ceramic scaffolds breaks down into three smaller packages:

- Find and establish suitable strategies for cell inoculation and cultivation on and in porous ceramics,

- establish qualified assay systems to verify vital growth of cells inside the ceramic matrix,

- chose appropriate scaffold geometries and design and operate a reactor device holding ceramic scaffolds (in strong cooperation with the departments mentioned above).

Progress of the work is outlined as follows: At first, cultivation surface, developed bioreactor device and operation techniques, used cell lines and employed analytical methods are introduced (chapter 3). After a short characterization of the utilized ceramics and flow inside the reactor system in chapter 4, two selected assay methods are discussed in more detail (chapter 5). A large part of the work is devoted to the development of an inoculation strategy which leads to good cell distribution inside ceramics, see chapter 6. This is followed by a demonstration of the reactor's feasibility as cell cultivation device in chapter 7 and 8 and proof of its suitablity for different cell types (chapter 9). The work closes with concluding remarks on accomplished goals and an outlook concerning open questions (chapter 10).

3 Materials and methods

3.1 Preparation of alumina ceramics for cell culture

Ceramics were supplied by H. Schubert, TU Berlin, whose working group consolidated the ceramic bodies as described in [Garrn et al., 2004]. As raw material highly purified aluminium oxide (AKP50, Sumitomo Chemical CO., LTD) was used, which was foamed by the help of protein (Albumin Bovine Fraction V, MP Biomedicals LLC) and dispersant Dispex A40 (Ciba Specialty Chemicals LTD). The resulting ceramics consisted of pore sizes mainly ranging from 150-200 µm with openings of 50-150 µm, porosity around 80-90% (see chapter 4.1.1) and specific surface of about $1\,m^2/g$, see figure 3.1.

All substances were milled for 15 min and consolidated in a microwave oven (µWaveVa0150, Püschner Mikrowellen Energietechnik) for 25 min at 400 W. Following drying in a drying closet (Memmert) for 15 min at 110°C, the ceramic bodies were cut to cylinders of 10 mm diameter using an auger. The protein scaffold was then removed at 600°C for 30 min in a muffle kiln (N20/HS, Naber Industrieofenbau) and ceramics were sintered (HT04/17, Nabertherm) at 1650°C for 60 min. Afterwards, the long foam cylinders were cut to cylinders of 5 mm or 10 mm height using a diamond saw. During this step, (mainly carbon-associated) tool residues that could interfere with subsequent analyzation methods can be deposited on the ceramic and, therefore, are removed by an additional torching step in the muffle kiln at 600°C for 30 min.

After this final purifying step, ceramics should be strictly prevented from any contact with a carbon source, being it proteins, fats, etc. Therefore, usage of gloves is strongly suggested.

To prepare ceramic cylinders for cell culture, foams were rinsed for 2 h in aqua dest., a detailed description of other not further specified substances and utensils is given in the Appendix, see page 101. Then, the lateral surface was covered in teflon tape (PTFE) and the foams were rinsed again to remove residual ceramic fragments. The teflon-sealing step ensures medium flow through the ceramic volume as the ceramic's flow resistance was found to be very high and medium otherwise tends to flow past the ceramic foam (see chapter 4.1.2). Foams were found to be robust against disruption by flow for at least seven weeks under continuous operation. Sterilization of the teflon-sealed ceramics was performed at 2 bar and 121 °C for at least 20 min.

3.2 Reactor device for cell cultivation on alumina foams

Allowing for initial demands discussed in chapter 2.1.1 and 2.1.2 and considering further requests as listed in table 3.1 a cultivation device was designed and constructed by Bischof

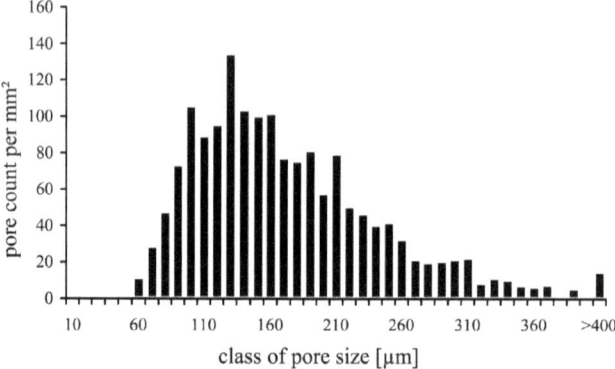

Figure 3.1: Pore size distribution for foamed alumina. Values were obtained by graphical analysis of scanning electron microscopy pictures (A. Berthold, TU Berlin). "Class of pore size" describes an interval of 10 µm below the listed index.

[Bischof and Blessing, 2006]. Considering all mandatory demands, the deduced prototype is a modular system consisting of standard subunits (further referred to as reactor modules) which can be easily converted to meet the needs of different operation strategies. Three main designs were manufactured which will be presented in more detail below. By providing nutrients out of one common medium reservoir, several reactor modules can be connected by one another either in line to simulate sequential reaction steps or side by side to adjust similar medium conditions, see figure 3.2a. By this, upscaling of the system is achieved rather by multiplication than volume enlargement, therefore, micro environments inside the porous scaffold can be maintanined. Moreover, during initial testing of foams, e. g., for new cell lines, foams with different characteristics such as porosity, thickness of the internal structures, pore volumes, etc. can easily be tested in one run using different modules or even within one module, see below. For all experiments described hereafter, reactor modules were connected individually to separate medium reservoirs if not stated otherwise, see figure 3.2b. Inside one reactor module, up to seven individual ceramic cylinders can be used in the actual design. They are either supplied with medium in series or in parallel as explained next.

3.2.1 Tubular design for series connection of ceramics

For series connection of ceramic foams, one to seven non-sterile foams (height 10 mm) otherwise prepared as in chapter 3.1 were assembled in a series reactor module made of poly-ether-etherketone (PEEK) and teflon (PTFE), which is connected to the medium reservoir by tubings. Medium perfusion is imprinted by a peristaltic pump. To separate foams from one another, seal rings (see Appendix) are used as spacers. A steel spring is used to hold foams in place if less than seven foams were used. The reactor unit is further equipped with inoculation ports made of aluminium and sight glasses (glass) to optically ensure bubble-free operation, see figure 3.3. For cultivation experiments, the reactor module was joined to a controlled bioreactor (Biostat

3. Materials and methods

Table 3.1: Demands for bioreactor design (as suggested by the author) at the starting point of reactor evolution.

mandatory	preferred	desired
autoclavable	easy to assemble	sterile disassembly
bubblefree operation	shakeable	addition of test ports
biologically inactive materials	easy to scale up	
parallel **or** series connection of single reactor units	continuous **or** pulsed medium perfusion	
laminar flow		
constant inlet process values (T, p, pO_2...)		
holds at least 6 ceramics at once		
sterile sampling		
sterile addition of supplements close to the foams		
incorporation of foams of different geometry		

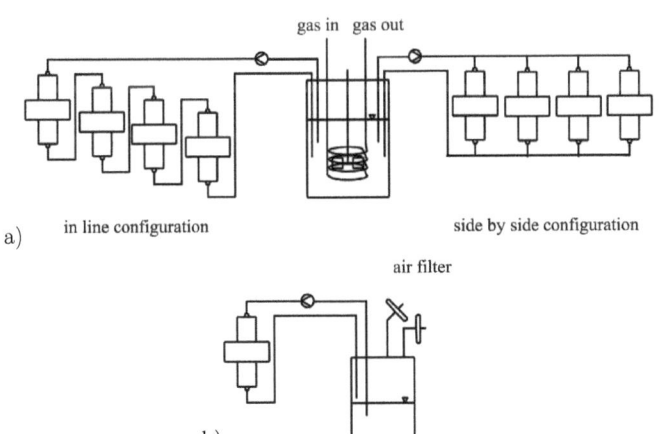

Figure 3.2: Module connection configurations. a) Connection of four reactor modules to one medium reservoir by an in line configuration (left) and side by side configuration (right). b) Testing configuration for individual reactor modules.

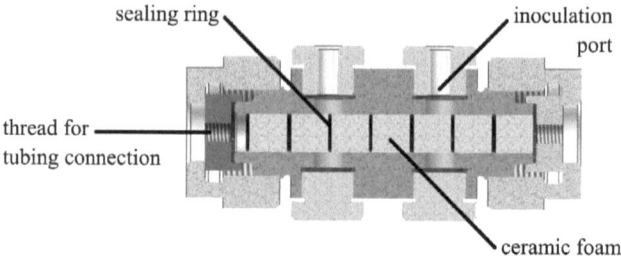

Figure 3.3: Tubular design reactor module for series connection of ceramic foams [Bischof and Blessing, 2006]. One to seven foams can be cultivated by perfusion at once.

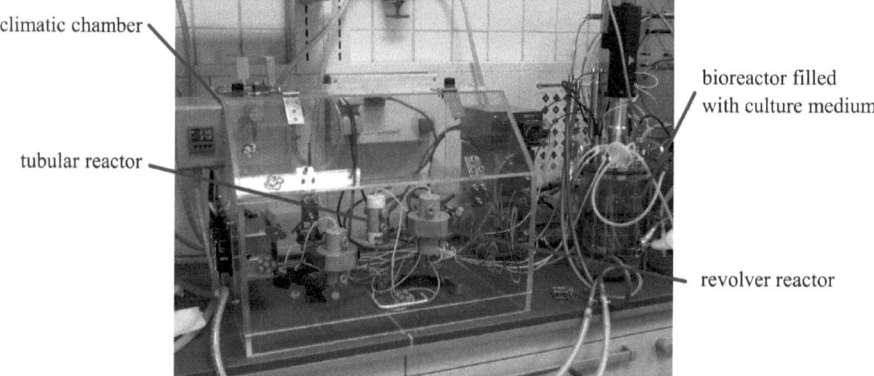

Figure 3.4: Connection of three reactor modules to a common medium reservoir. "Tubular reactor" refers to the design described in chapter 3.2.1, "revolver reactor" refers to the design described in chapter 3.2.2.

B plus, Sartorius), which was prepared for cell cultivation and equipped with a gas permeable tube. The system was then autoclaved as one unit, the bioreactor and the reactor module were filled with medium and the reactor was put in a climatic chamber for temperature controlled cultivation, see figure 3.4.

3.2.2 Revolver design for parallel connection of ceramics

For parallel connection of ceramic foams, six or seven non-sterile ceramic foams otherwise prepared as in chapter 3.1 were assembled in a revolver reactor module via a foam holding magazine made of PEEK which was selected depending on the foams outer geometry and the inoculation strategy pursued, see figure 3.5. The magazine is embedded between two PEEK shells and sealed by a seal ring and a silicone sealing disk. The shell's and magazine's geometry was continuously adapted during the course of the project. Two versions will be used in this work, see figure 3.6. The module is bolted by an aluminium throw and connected by tubings either to a controlled bioreactor (Biostat B plus) or to a medium reservoir that can be put

Figure 3.5: Ceramic foam holding magazines for static inoculation (plain magazine, upper row) and dynamic perfusion inoculation (cone magazine, lower row) [Bischof and Blessing, 2006]. Two depth of the six bowls for the lower design are available: 5 mm and 10 mm. For cultivation, perfusion is imprinted from above (second column).

into a CO_2-incubator (see figure 3.2). Medium perfusion then is imprinted by a peristaltic pump operated outside the incubator. The reactor module is further equipped with inoculation and sampling ports (and if desired with sight glasses to ensure bubble-free operation). For a complete sketch of the two main reactor designs see figure 3.6. The system was autoclaved prior to usage and residual water was removed before the module was filled completely with culture medium.

3.2.3 Small block design for single foam analysis

For some experiments, it was beneficial to analyze a single foam at once. Therefore, a small block capable of holding one ceramic foam was designed, see figure 3.7. The reactor module consists of two shells which are bolt together and hold a sealing ring with a teflon sealed ceramic foam plugged in. The module is connected to a medium reservoir by tubings and operated in a CO_2 incubator. Due to the simple design and therefore lack of ports, the foam had to be inoculated with cells outside the reactor device (chapter 3.3.1) and then aseptically assembled into the separately sterilized module. Filling of the module and perfusion with culture medium is performed by help of a peristaltic pump as above.

3.3 Inoculation procedures

3.3.1 Static inoculation

For some experiments static inoculation of cells was inevitable and performed as follows: Following preparation of ceramics as in chapter 3.1, foams were soaked in culture medium for at least 2 h at 37°C. I hereby expect the ceramic's surface to absorb serum proteins of the medium and therefore cause a beneficial ground for cell attachment. Cell inoculation is then performed from above, whereby cells (suspended in pre-warmed medium) are dropped directly onto the foam's surface in 100-200 µl volume at the maximum. Following a 30 min adherence

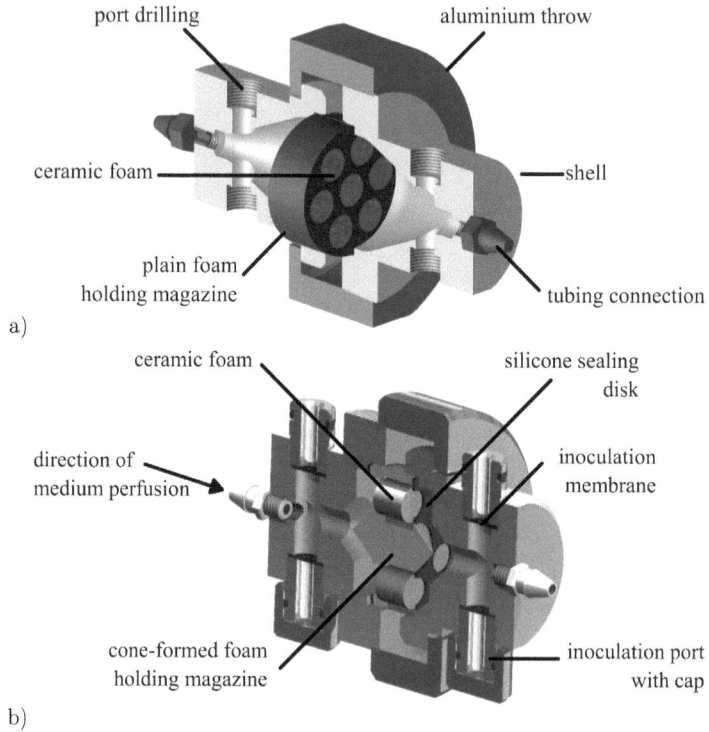

Figure 3.6: Reactor modules for parallel perfusion of several ceramic foams [Bischof and Blessing, 2006]. a) Early design with conically, flow-calming zones before and after the foam holding magazine for potentially turbulent flow velocities. b) Final design with less void volume and double-cone magazine to direct medium flow uniformly over all foams (compare to figure 3.5).

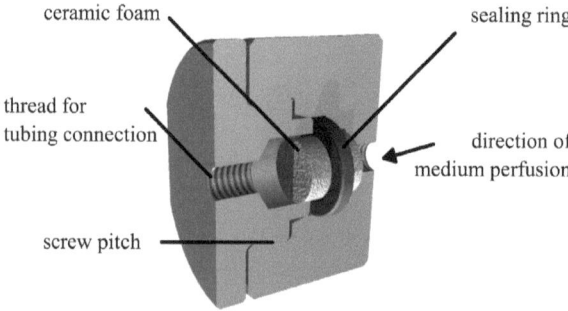

Figure 3.7: Sketch of a reactor module for cultivation and/or analysis of a single foam [Bischof and Blessing, 2006].

time, whose adequacy in terms of strong attachment of cells on substrate was confirmed experimentally (data not shown), experimental procedures continued.

If static inoculation in the bioreactor module was to be performed, sterile reactor modules filled with medium and pre-warmed at 37°C inside a climatic chamber or a CO_2-incubator were inoculated with cells via a needle and syringe from above, the module was slightly shaken for better cellular distribution, and cells were allowed to settle for 2-4 h before cultivation proceeded.

3.3.2 Dynamic inoculation by stirring or agitation

For dynamic inoculation, unsealed ceramic foams with 5 mm height and a concentrical borehole of 2 mm diameter otherwise prepared as in chapter 3.1 were hitched into a 50 ml reaction tube with a flexible steel wire. Tubes were filled with 15 ml cell suspension in culture medium and immediately moved back and forth at 37°C on an agitator (self-build, suitable for use in a CO_2 incubator) for 72 h at 1/s.

3.3.3 Dynamic inoculation by convectional forces

For perfusion assisted inoculation, revolver reactor units with conical foam holding magazines as described in chapter 3.2.2 were further equipped with additional components prior to sterilization, see figure 3.8. Modules were placed in the CO_2 incubator and filled with medium. Perfusion was performed overnight to ensure adjustment of temperature (37°C) and pH-value of medium. An upper glass olive was filled with an extra medium volume of 10-20 ml and cells were inoculated in 500 µl medium via the upper inoculation port. In order to degas foams thoroughly, internal pressure was reduced for 30 s by removing 1 ml medium out of the closed system via the lower sampling port and passing it back into it. The extra medium was then pumped from the upper glass olive to the lower, passing over the ceramic foams and flushing cells throughout the foam volume. Pumping medium back and forth was repeated several times with one pass of cells over the foam corresponding to 0.5 cycles. During dynamic seeding, modules normally were aligned vertically. However, for some experiments, modules were aligned horizontally and turned for 90° around the horizontal axis at every 0.5 cycles.

For another experimental setup, modules were aligned vertically and held horizontally every 0.5 cycles for 90 s with stopped pumps, rotating the horizontal alignment by 90° for every half cycle.

After dynamic inoculation, cells were allowed to adhere for 30 min before perfusion started from above.

3.4 Cultivation of cells on ceramics

The development and optimization of a perfused bioreactor system based on ceramic foams was the main focus of the TU Berlin funded project IFS3/2, see chapter 2.2. Therefore, perfusion cultivation of cells on ceramic foams was performed in the reactor devices described above.

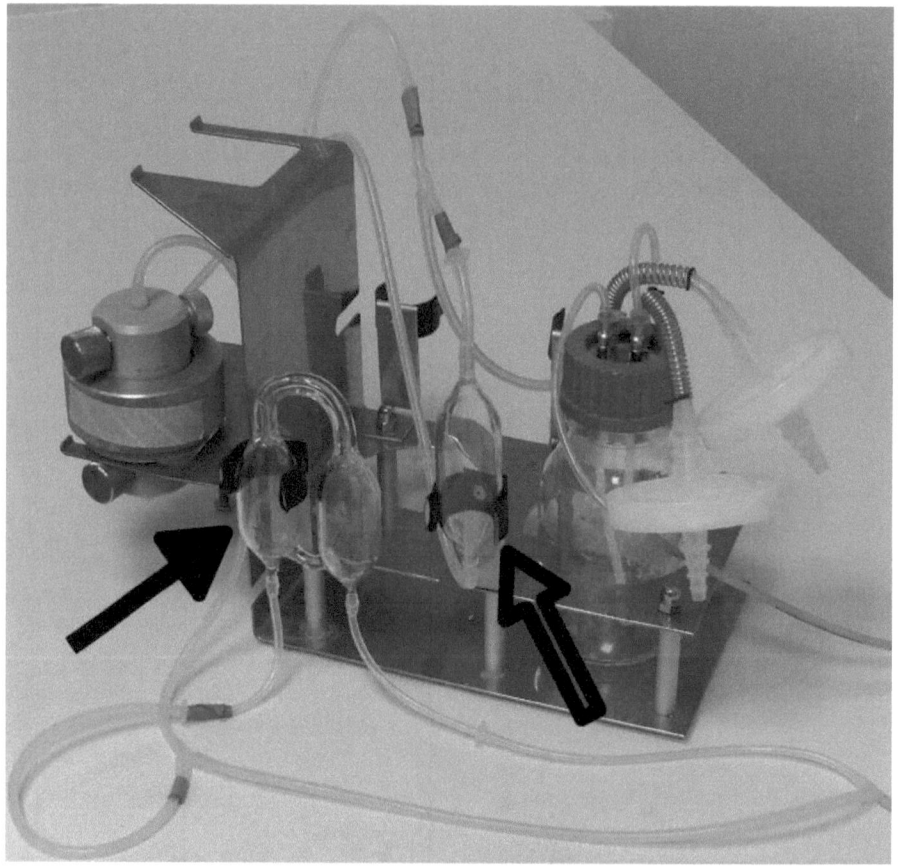

Figure 3.8: Setup of a reactor module and medium reservoir additionally equipped with glass olives for dynamic inoculation in a CO_2 incubator. The module (left) comprises a total volume of about 20 ml, reservoir (right) contains 100-200 ml culture medium. Medium is pumped through the modules via silicone tubes by peristaltic pumps, which are operated outside the incubator. The filled arrow points to an upper glass olive, the blank one to a lower olive.

However, for some experiments, cultivation of cells was performed outside reactor devices, leading to different operation strategies as follows.

3.4.1 Perfusion cultivation inside the reactor device

Following attachment, cells on ceramics in reactor devices were cultivated under perfusion conditions, i.e. medium was pumped through ceramics. Depending on visual evaluation of the medium's color in the medium reservoir, which is orange for pH=7.2 and proceeds to bright yellow for pH=6 (due to pH-indicator phenole red), medium was partially replaced at $pH \lesssim 6.5$ to prevent accumulation of lactate. In doing so, the medium containing reservoir was replaced with another vessel containing fresh medium, which had been equilibrated in the CO_2-incubator for at least 4 h. This procedure took about 10 min, therefore perfusion cultivation was interrupted for 10 min at the longest. Sampling of the reactor module medium was performed via the lower inoculation port, meaning medium samples of 1 ml were taken from behind the ceramics in flow direction.

3.4.2 Static cultivation of cells outside the reactor device

For static cultivation, cell populated foams were transferred to empty culture vessels and covered with pre-warmed medium. Depending on optical evaluation of the medium's color, medium was replaced at $pH \lesssim 6.5$ to prevent accumulation of lactate. In doing so, all medium was discarded and replaced by fresh medium, which had been equilibrated in the CO_2-incubator for 4 h.

3.4.3 Cultivation of cells by medium convection outside the reactor device

For convective cultivation of cells, cell-populated foams with a concentrical borehole were hitched into 50 ml reaction tubes containing 15 ml pre-warmed culture medium by a flexible steel wire. Tubes were placed on an agitator (self-build, suitable for use in a CO_2 incubator) and moved back and forth at 1/s and 37°C.

3.5 Cell lines and origin

For all experiments identifying operation strategies and parameters as well as scaffold geometries, which support cellular growth best, I used the mammalian cell line *CHO-K1* (chinese hamster ovary, provided by U. Reichl, MPI Magdeburg) due to its decade-long prominence in scientific literature and its rather simple handling techniques. Moreover, data regarding *CHO* cell metabolism is widely accessible [Wu et al., 1992, Hansen and Emborg, 1994, Altamirano et al., 2001, Deshpande and Heinzle, 2004] and therefore interpretation of cultivation experiments is facilitated. A significant body of work was done to adapt the *CHO-K1* cell line to growth in medium with very low serum concentrations and to experiments on ceramic

Table 3.2: Cultivation requirements for *CHO-K1, A549, MDCK* and human fibroblasts. * refers to cells adapted to growth in medium containing 1% fetal bovine serum, see Appendix.

cell	medium	average population doubling time (as observed in experiments)
CHO-K1	90% Ham-F12 (Sigma N4888), 10% FBS (Gibco, 10270-106, Lot 41G6972K), 2 mmol/l glutamine (Merck, 1.00289)	17 hours
CHO-K1, 1% FBS*	99% Ham-F12 (Sigma N4888), 1% FBS (Gibco, 10270-098, Lot 41G1360K), 2 mmol/l glutamine (Merck, 1.00289)	17 hours
A549	90% DMEM-Low Glucose (Sigma D5523), 10% FBS (Gibco, 10270-106, Lot 41G6972K), 3,7 g/l sodium bicarbonate (Merck, 1.06329)	70 hours
MDCK	90% Ham-F12 (Sigma N4888), 10% FBS (Gibco, 10270-106, Lot 41G6972K), 2 mmol/l glutamine (Merck, 1.00289)	17 hours
Fibroblasts	90% DMEM+GlutaMAX (Gibco 61965), 10% FBS (Gibco, 10270-106, Lot 41G6972K)	3 weeks

foams with these adapted cells. However, as growth on ceramics was poor, e.g., for a serum content of 1%, all but one set of experiments in this report are based on medium with 10% fetal bovine serum. More details relating to the serum-deprived cells can be found in the Appendix. To prove the generality of the established bioreactor system along with the elaborated operation modes, I furthermore introduced three more cells: *A549* is a lung carcinoma cell line [Lieber et al., 1976] widely used in toxicity-testing [Narula et al., 1998, Davoren et al., 2006] and was gently provided by A. Hartwig, TU Berlin. *MDCK* (Madin-Darby canine kidney) is a mammalian cell line known for efficient virus production [Genzel et al., 2004, Möhler et al., 2008] and was gently provided by U. Reichl. Cells were adapted to growth in Ham-F12 medium prior to usage in experiments. Human primary fibroblasts from small skin biopsies were gently provided by R. Lauster, TU Berlin. Table 3.2 summarizes the requirements for cell cultivation for all cell types. Cultivation experiments and metabolic assays as in chapter 3.6.4 and 3.6.5 were performed at 37°C and a CO_2 atmosphere of 5% for all cells. Cells were used for experiments in their exponentially growth phase, unless otherwise stated. Cultivations preceeding described experiments were performed according to the Appendix.

3.6 Assays

To evaluate the qualification of a cultivation device, it is necessary to analyze cell number and cell distribution on the underlying scaffold. Common attempts to detach cells and count them by standard techniques are not easily transferred to porous structures and failed. As many scaffolds consist of a cell entrapping structure similar to that of a porous ceramic, recently other approaches were published, among them counting detached nuclei [Lee et al., 1991] and measuring dissolved DNA (Cyquant® Cell Proliferation Assay Kit, Invitrogen). Attempts to adapt these methods to the system at hand, however, failed (more detailed results can be found in [Goralczyk et al., 2009]). I therefore established an entirely new approach of estimating biomass on and in the scaffold by carrier hot gas extraction, see below. Moreover, I indirectly assessed cell number by evaluating the metabolic status of cells grown on ceramics and compare it to those of cells grown in culture flasks, hereby measuring culture medium glucose and lactate concentration and performing a metabolic assay based on the reduction of resazurin, see below. I am aware, of course, that such a metabolic evaluation gives only a very rough estimate, as factors such as inhomogeneous nutrient supply, time-varying metabolism, etc. strongly affect the results obtained.

To visually validate cellular distribution on and inside the ceramic scaffold, I adapted two dyeing protocols as stated below.

3.6.1 Microscopic evaluation of cell vitality by dyeing according to FDA/EB protocol

Fluorescein diacetate (FDA) and ethidium bromide (EB) are often used dyes to differentiate live and dead cells in cell cultivation. Hereby, colorless FDA on the one hand is incorporated into cells and intracellularly hydrolyzed to fluorescein, which fluoresces green when excited with blue light. This reaction being catalyzed by active esterases only takes place in alive, metabolic competent cells.

EB on the other hand can only invade dead cells as it only passes disintegrated cell membranes. Inside the cell, it binds to DNA and fluoresces red. The combination of both dyes consequently allows a simple classification of a cell's vital status.

Statically cultured foams were rinsed with prewarmed PBS++ (37°C, see appendix, chapter 10) and gently wiped with paper towels to remove liquids. If ceramics were cultured in a reactor device, modules were rinsed with 80 ml prewarmed PBS++ for 30 min and foams were removed from the reactor magazine and gently wiped with paper towels. Foams were then soaked in a 5 µg ethidium bromide (Sigma-Aldrich) and 25 µg fluorescein diacetate (Sigma) physiological salt solution (FDA/EB staining). Cells were differentiated by fluorescent microscopy (Axioscope, Zeiss), whereby live cells fluoresce green and dead cells red. Digital images were assembled and annotated using PaintShop Pro and Flash.

To include cells growing inside the scaffold into analysis, foams were cut with a scalpel blade and cross sections were also analyzed as described.

3.6.2 Microscopic evaluation of cell distribution by dyeing according to hematoxylin/eosin protocol

Dyeing according to a hematoxylin/eosin protocol differentiates almost all cells and tissues into cellular background and nuclei. Following dyeing procedure, cells appear in shades of pink and nuclei appear blue.

Statically cultured foams were rinsed three times with prewarmed PBS++ and gently wiped with paper towels. If ceramics were cultured in a reactor device, modules were rinsed with 80 ml prewarmed PBS++ for 30 min and foams were removed from the reactor magazine and gently wiped with paper towels. Foams were then fixed in ice-cold methanol for 10 min and thoroughly dried under the hood before storing at -80°C.

For dyeing, foams were washed twice in aqua dest for 2 min each, then stained with Harris Hematoxylin (Chroma, 2C 165) for 1 min. Cells were blued under tap water for 1 min and cross-stained in 1% Eosin (Chroma, 2C 284) for 1 min. Following a washing step in aqua dest for 1 min and dehydration in 70%, 96% and 100% ethanol for 2 min each, foams were thoroughly dried under the hood and cells were differentiated by light microscopy (Axioscope, Zeiss). Digital images were assembled and annotated using PaintShop Pro and Flash.

To include cells growing inside the scaffold into analysis, foams were cut with a scalpel blade and cross sections were also analyzed as described.

3.6.3 Scanning electron microscopy of ceramic surface and cells on ceramics

For scanning electron microscopy (SEM), following cultivation, ceramic foams were washed twice in prewarmed PBS++, then cells were fixed in Karnovsky buffer overnight at 4°C. Then, samples were washed twice in cacodylate buffer and dehydrated in solutions containing increasing percentages of acetone (10%, 30%, 50%, 70%, 90%) for 15 min each and four times 15 min each in 100% acetone [Anton et al., 2008]. Dry foams were imaged with an XL20 (Philips) and images further assembled and annotated using PaintShop Pro and Flash.

3.6.4 Metabolic evaluation of glucose consumption and lactate formation

Medium was analyzed for the concentration of glucose (Glucose GOD FS, DiaSys) and lactate (Enzytec fluid L-lactate, DiaSys). Whereas manufacturers specify limits of quantification to be 0.01 g/l for glucose and 0.003 g/l for lactate, at laboratory performance (data not shown) 95% confidence interval $CI_{95\%}$ for measurement of glucose concentration $c_{glucose}$ encompassed $\pm 0.12 * c_{glucose}$ for concentrations of 0.25 g/l and higher. Regarding measurements of lactate concentration $c_{lactate}$, $CI_{95\%}$ encompassed $\pm 0.2 * c_{lactate}$ for concentrations of 0.15 g/l and higher. For lower concentrations confidence intervals were at $\pm 0.18 * c_{glucose}$ for glucose and $\pm 1 * c_{lactate}$ for lactate.

For cultivation in reactor modules, accumulated glucose consumption and lactate formation

were calculated as follows.
Glucose consumption:

$$m_{glucose}(t_n) = m_{glucose}(t_{n-1}) + \Delta m_{glucose}(t_n, t_{n-1}) \tag{3.1}$$

$$\Delta m_{glucose}(t_n, t_{n-1}) = V_{system} * (c_{glucose}(t_{n-1}) - c_{glucose}(t_n)) \tag{3.2}$$

$$m_{glucose}(t_0) = 0 \tag{3.3}$$

Lactate formation:

$$m_{lactate}(t_n) = m_{lactate}(t_{n-1}) + \Delta m_{lactate}(t_n, t_{n-1}) \tag{3.4}$$

$$\Delta m_{lactate}(t_n, t_{n-1}) = V_{system} * (c_{lactate}(t_n) - c_{lactate}(t_{n-1})) \tag{3.5}$$

$$m_{lactate}(t_0) = 0, \tag{3.6}$$

where m_x is the accumulated mass of species x, c_x is the concentration of species x in culture medium, t_n is a discrete time point of sampling and V_{system} is the total medium volume recirculating over the module. To simplify the calculation, at each time point medium concentration in the reservoir was assumed to equal medium concentration in the reactor module behind the ceramic foams. Being clearly an incorrect assumption, it still enables a simple comparison of different test runs.

3.6.5 Reduction of resazurin

Simple metabolic assays as MTT, XTT [Scudiero et al., 1988] or Alamar Blue [Ahmed et al., 1994] are long time known and well characterized. A quite simple method to evaluate the metabolic status of a cell culture is to monitor the reduction of resazurin (which by the way is the main component in Alamar Blue [O'Brien et al., 2000]). Resazurin is a blue, "nonradioactive, nontoxic, water-soluble (eliminating the need for extraction) and readily detectable by either absorbance or fluorescence spectroscopy" [Voytik-Harbin et al., 1998] dye whose color turns into pink when reduced to resorufin and which becomes colorless when further reduced to hydroresorufin. For resazurin reduction up to 50%, [Anoopkumar-Dukie et al., 2005] found a linear correlation to cell number, whereby reduction occurs by mitochondrial, cytosolic and microsomal enzymes [Gonzalez and Tarloff, 2001].

Resazurin assays were performed either inside the reactor module itself or for single foams in 12-well culture plates. For assays in the reactor device, medium was discarded and the module was rinsed with 80 ml prewarmed PBS++ for 30 min. PBS++ was discarded and the module was filled with 40 ml pre-equilibrated culture medium containing 22 µmol/l resazurin (Sigma). Perfusion proceeded and medium samples were taken at discrete times from the lower inoculation port and photometrically analyzed (Tecan Sunrise, Tecan Trading AG) at 570 and 600 nm. Percentage of reduced resazurin was calculated as in equation 3.7 (provided in Alamar

Blue datasheet, Serotec).

$$p = 100 * \frac{117.216 * AS_{570nm} - 80.586 * AS_{600nm}}{155.677 * AB_{600nm} - 14.652 * AB_{570nm}}, \qquad (3.7)$$

where p is the percentage of reduced resazurin (here resorufin), $AS_{wavelength}$ represents the absorbance of the sample at a given wavelength and $AB_{wavelength}$ that of a negative control.

For resazurin assays in small block reactors as described in chapter 3.2.3, following sterile assembly of populated ceramics into the small block reactor, the reactor was filled with 6 ml (V_{system}) pre-equilibrated culture medium containing 22 µmol/l resazurin. Perfusion proceeded for 165 min and medium was collected and photometrically analyzed as above. Percentage of resorufin was calculated as in equation 3.7 and accumulation of resorufin was calculated as in equation 3.8.

$$m_{resorufin} = V_{system} * c_{sum} * \frac{p - p_0}{100} \qquad (3.8)$$

$$c_{sum} = 5\mu g/ml, \qquad (3.9)$$

where $m_{resorufin}$ is the mass of reduced resazurin, V_{system} is total medium volume in which the reduction takes place, p and p_0 are percentage of resorufin at the end of the assay and its beginning, respectively. As percentage of resorufin is not zero at $t = t_0$ (as calculated from equation 3.7), resazurin in batch medium is a mixture of both resazurin and resorufin (corresponding to c_{sum}).

For resazurin assays for cells on ceramics performed in 12- or 24-well culture plates, foams were covered with pre-equilibrated culture medium containing 22 µmol/l resazurin. Following static cultivation, medium was vigorously pipetted up and down and samples were analyzed for resazurin reduction as above (equation 3.8).

For dynamic resazurin assays for cells statically cultivated in culture flasks or culture wells, cells were covered with culture medium containing 22 µmol/l resazurin. Medium samples were taken in intervals and cumulative reduction of resazurin was calculated as in equation 3.10.

$$m_{resorufin}(t_n) = m_{resorufin}(t_{n-1}) + \Delta m_{resorufin}(t_{n-1}, t_n) \qquad (3.10)$$

$$\Delta m_{resorufin}(t_{n-1}, t_n) = V_{system}(t_n) * c_{sum} * \frac{1}{100}(p(t_n) - p(t_{n-1})) \qquad (3.11)$$

$$V_{system}(t_n) = V_{system}(t_{n-1}) - V_{sample} \qquad (3.12)$$

where $m_{resorufin}(t_n)$ is the accumulated mass of reduced resazurin (resorufin) at a given sampling time point n, $V_{system}(t_n)$ is the total medium volume in which the reaction takes place for each time interval $[t_{n-1}, t_n]$ (and which is reduced during the assay due to sampling) and $p(t_n)$ is the percentage of resorufin at a given time point as calculated in equation 3.7.

3.6.6 Carrier hot gas extraction

For a first approximation of the existing biomass on and in a ceramic scaffold, I developed a very new method, further referred to as carrier hot gas extraction (CHGE), in collaboration with O. Goerke from TU Berlin. The principle relies on the coupling of biomass to mass of elementary carbon, which can be determined by incinerating ceramics completely and analyzing combustion gases, in particular CO_2.

Following cultivation, ceramic foams were washed in prewarmed PBS++ and dried in a laboratory oven at 50°C overnight before storage at -80°C. For carrier hot gas extraction, foams were dried at 50°C, then crushed to pieces between two layers of thin tin foil (Carl Roth) in a Zwick Z00z (Zwick GmbH & Co. KG, Ulm) and thoroughly dried at 50°C overnight. Ceramic pieces were put in ceramic cups (IVA-Analysetechnik e. K.) which were cauterized prior to use in order to ensure total removement of carbon. For even heat distribution during the incinerating process, 0.3 g tin (IVA-Analysetechnik e. K.) and 3 g tungsten (IVA-Analysetechnik e. K.) were added and cups were placed in a Carbon/Sulfur Elemental Metal Infrared Analyzer (EMIA-320V, Horbia). Incineration occurred at 2500-3000°C. Combustion gas was analyzed simultaneously for carbon dioxide and sulfur dioxide, which was computed as percentage of the ceramic pieces' total weight. Calculation of elementary carbon mass was performed following equation 3.13.

$$m_{carbon} = \frac{p_{carbon}}{100} * m_{ceramic}, \qquad (3.13)$$

where m_{carbon} equals the total mass of elementary carbon, p_{carbon} is the percentage of carbon as determined in CHGE and $m_{ceramic}$ is the total mass of the ceramic pieces. Calculation of elementary sulfur mass was performed analogously.

4 Reactor characterization

4.1 Ceramics

4.1.1 Porosity

Porosity of ceramics was calculated to be $\phi = 0.87$ with a standard deviation of $\sigma = 0.005$ following equation 4.1. Hereby, three dry ceramic cylinders of 20 mm height and 25 mm diameter each (corresponding to an outer ceramic volume V_{total} of 9.8 ml) were considered for determination.

$$\phi = \frac{V_{void}}{V_{total}} = \frac{V_{total} - V_{Al_2O_3}}{V_{total}} \tag{4.1}$$

$$V_{Al_2O_3} = \frac{m_{Al_2O_3}}{\rho_{Al_2O_3}} \tag{4.2}$$

Variables correspond to: V_{total} outer ceramic volume, $V_{Al_2O_3}$ volume of ceramic material, $m_{Al_2O_3}$ mass of ceramic material as determined by weighing (HM-200-EC, A&D Instruments Ltd) and $\rho_{Al_2O_3}$ density of pure alumina which is 3.98 g/ml.

In order to determine pore distribution, scanning electron microscopic images from the top of one selected ceramic were analyzed for nominal pore count (ppi, pores per inch). Six images were evaluated and revealed a ppi of 135, corresponding to a mean pore diameter of 190 μm. For this pore size, porosity for porous ceramics is around $\phi = 0.9$ [Innocentini et al., 1998] which is in good agreement with experimental results.

4.1.2 Flow resistance

As a scale for flow resistance, I evaluated Darcy's constant K_d whereby a lower value of K_d indicates a higher flow resistance. For a sketch of the experimental setup see figure 4.1. Ceramics of 15 mm height and 10 mm diameter were fitted in a heat shrinking tube (SDH 19 SW, Reichelt Elektronik) of approximately 30 mm length and assembled horizontally between two silicone pipes. Water was pumped through the construct whereby volumetric flow was adjusted with a branching drain valve. Overpressure upstream of the ceramic was evaluated from height of water column in a vertical branch of the pipe ahead of the ceramic and pressure downstream of the ceramic was assumed to equal atmospheric pressure due to the open setup. Water was collected at the end of the pipe and weighed for time intervals of 30 s and volumetric flow velocity was calculated according to equation 4.4. Following equation 4.3 Darcy's constant (i.e. darcian permeability) was estimated to be about $K_d = 1.7 \cdot 10^{-10}\,\text{m}^2$ with a standard deviation of

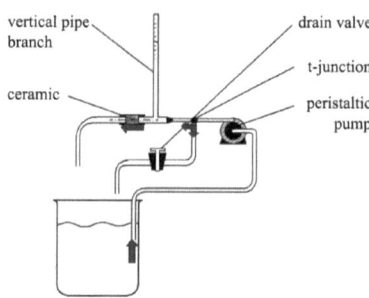

Figure 4.1: Experimental setup for evaluation of K_d. Pressure drop is evaluated from water column as read in vertical pipe branch. Direction of water flow is indicated by blue arrows.

$\sigma = 7.1 \cdot 10^{-12}\,\mathrm{m}^2$ for 10 measurement points. Permeability therewith is rather low compared to ceramics with a higher porosity or lower ppi as summarized in [Innocentini et al., 1998]. I therefore decided to seal ceramics laterally for perfusion experiments, as a high flow resistance otherwise could lead to medium flowing by the ceramics.

$$K_d = \frac{\dot{V} * \eta * H}{A * \Delta p} \quad (4.3)$$

$$\dot{V} = \frac{m_{flow}}{\rho_{H_2O} * 30\,\mathrm{s}} \quad (4.4)$$

Variables correspond to: $\dot{V}$ volumetric flow velocity, η medium coefficient of viscosity, H scaffold height in direction of perfusion. A perfused scaffold area, Δp pressure drop across perfused scaffold, m_{flow} water throughput for 30 s and $\rho_{H_2O} = 998\,\mathrm{kg/m^3}$.

To demonstrate efficiency of lateral foam sealing, in another experimental setup, ceramics of 20 mm height and 25 mm diameter were assembled into vertically aligned silicone pipes with variations in sealing, see table 4.1. Water was pumped through the pipes with a constant water column above the ceramics and a constant drain of around 50 ml/min. For direct optical evaluation of flow through the ceramic, red aqueous ink was applied directly on top and in the middle of the ceramic via a long needle and syringe and the arising flow filament was observed and documented, see table 4.1, right column.

For evaluation of flow distribution inside the ceramics, ceramics were coated with 0.25% bovine serum albumine (Merck, 1.12018) in aqua dest. at 37°C overnight directly before experimental procedures. Here, 10% coomassie blue (Fluka, 27813) in aqua dest. was used for perfusion of ceramics. Following 5 to 10 min of perfusion, ceramics were dismantled and cut vertically for determination of color distribution, see table 4.1 middle column, blue color corresponds to flow having past through the analyzed volume element as proteins bind the coomassie dye.

Both ink and coomassie blue distribution showed a homogenous liquid throughput through the total ceramic volume only for thorough lateral sealing, last row in table 4.1. Therefore lateral sealing is strongly advised if homogeneous flow distribution is sought.

4.2 Characterization of flow inside the reactor

A simple way to characterize a perfusion reactor is to compare its flow profile to that of an idealized stirred tank reactor (STR) and that of an ideal plug flow reactor (PFR). Here, Hagen [Hagen, 2005] considers the dimensionless Bodenstein number Bo as a measure for the ratio of enforced convection to dispersion which is zero for an ideal STR and converges to infinity for an ideal PFR. From that, the axial coefficient of dispersion D_{ax} in m^2/s is calculated following equation 4.5.

$$D_{ax} = \frac{u * L}{Bo} \tag{4.5}$$

$$u = \frac{L * \dot{V}}{V}, \tag{4.6}$$

where L is the length of the reactor, $\dot{V}$ is the volumetric flow velocity, V is the total reactor volume and Bo is the Bodenstein number as deduced from step experiments.

The Bodenstein number was determined for the revolver design reactor of chapter 3.2.2 for both a plain foam holding magazine and a double conical magazine as in figure 3.5. In order to obtain comparable results, for a third experiment the bowl in the middle of the plain magazine was clogged.

Step experiments were performed as follows: Modules were assembled as for cultivation with 6 or 7 ceramic foams of 10 mm height as in chapter 3.2.2 and filled with aqua dest. Perfusion was performed from above at 1.4 ml min^{-1}. As a tracer, 1 mg/ml resazurin in water was joined at time t_0 and samples of the flow off were analyzed photometrically at 600 nm. For evaluation, time was normalized to dimensionless time θ via equation 4.7 and extinction was normalized to dimensionless $F(\theta)$, which is the function of sum of retention time via equation 4.9.

$$\theta = \frac{t}{t_{eff}} \tag{4.7}$$

$$t_{eff} = \sum_{n=1}^{m} (1 - F(t_n)) * \Delta t_{n-1,n} \tag{4.8}$$

$$F(\theta) = \frac{E(\theta)}{E_\infty}, \tag{4.9}$$

where t_n is a time point of sampling, t_m is the time point of sampling when $E(t_n)$ equals E_∞, $E(\theta)$ is the fluid extinction at each normalized time point and E_∞ is the maximal extinction reached in the experiment. $F(t_n)$ is calculated analogously to equation 4.9.

Figure 4.2 shows the experimental results for reactor characterization.

Bodenstein number and coefficient of axial dispersion were calculated as in equation 4.10 and

Table 4.1: Flow distribution inside ceramics for different sealing strategies, see text for further explanation. Blue arrows indicate direction of flow; a deeper color for a volume element indicates higher throughput for that area.

sealing setup	coomassie blue staining	evaluation of ink flow filament
ceramic hooked inside pipe, no lateral sealing	dye is found solely on top of the ceramic with poor invasion in the middle	ink mainly passes by the ceramic
ceramic hooked inside pipe, lateral sealing at top of ceramic	homogeneous distribution of dye at top of ceramic	ink flows laterally into the ceramic and leaves it directly behind the sealing
ceramic hooked inside pipe, lateral sealing at bottom of ceramic	inhomogeneous distribution of dye	ink flows directly into the ceramic and escapes all over the bottom of the ceramic
ceramic grafted into pipe, lateral sealing with teflon tape	homogenous distribution of dye all over the ceramic's volume	ink flows directly into the ceramic and escapes all over the bottom of the ceramic

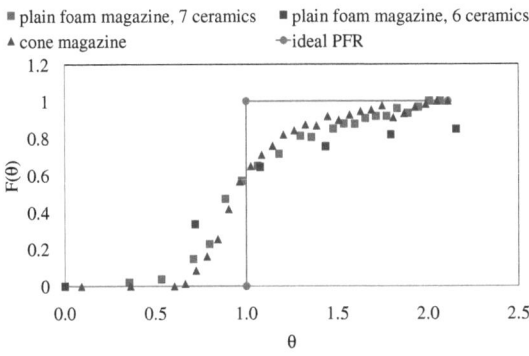

Figure 4.2: Color step experiments for different foam holding magazine geometries (data from one experiment per reactor design). Shown is normalized extinction $F(\theta)$ over normalized time θ.

Table 4.2: Reactor characteristics for plain and cone magazine holding 6 or 7 ceramic foams.

magazine geometry	slope	Bo	D_{ax} [m²/s]
plain, 7 foams	0.9	10	0.00057
plain, 6 foams	0.8	9	0.00065
conical	1.1	16	0.0003

equation 4.5 and are listed in table 4.2.

$$Bo = \pi * (2 * a)^2, \tag{4.10}$$

where a is the slope of the graph calculated from two points around $\theta = 1$.

Comparing flow for 6 and 7 ceramics in a plain holding magazine shows that a set-up of seven ceramics resembles more an ideal PFR, see figure 4.2. Therefore, it is assumed that the open structure with a flow path in the middle of the magazine maintains a more laminar flow profile than for a magazine with a blocked flow path in the middle. However, the difference between 6 and 7 integrated ceramics is not very high, see table 4.2.

Clearly, the shape of the foam holding magazines as well as the shape of the magazine shells influences the perfusion flow profile. A conical magazine with narrow flow entry resembles more a PFR as the Bodenstein number is higher and the color profile in figure 4.2 is closer to the ideal profile than for a plain magazin in a reactor with a widening shell. Nevertheless, the difference between both is rather small, and therefore I rejected the widening shell design for the narrow one in favor of the material savings.

Regarding the magazine's shape, I performed a similar experiment as described above with a tracer dye sticking to the ceramics. I hereby proved that flow mainly passes through the foam in the middle of the plain magazine for the 7-foam assembly - as optical evaluation revealed a strong staining of the middle foam towards the ceramics at the rim (data not show). Therefore, during the course of the presented work, I focused on employing the conical magazine to get a homogenous cell distribution and medium supply among all foams.

5 Coupling of cell number to metabolic reduction of resazurin and to elementary carbon analyzed in CHGE

5.1 Reduction of resazurin by *CHO-K1*

5.1.1 Determination of rate of reduction

In order to evaluate the applicability of the resazurin assay for determination of the *CHO-K1* cell number, two dimensional experiments were performed in cell culture flasks and plates. Different cell numbers (passage 3) were inoculated in culture medium onto flasks or plates with varying surface area to modulate cell densities around $4 \cdot 10^4$ cells/cm^2, which are assumed to be exponentially growing, and cell densities around $8 \cdot 10^4$ cells/cm^2 or higher, which are assumed to be in a stationary phase (boundaries were obtained in earlier experiments, data not shown). Cells were allowed to adhere for 2 h, then medium was discarded, cells were washed with PBS++ and covered with medium containing 22 µmol/l resazurin. Medium volumes were set to be 20 ml for flasks with a growth surface of 175 cm^2, 10 ml for 75 cm^2 flasks, 6 ml for 25 cm^2 flasks and 5 ml for 6-well plates with a growth surface of 9.6 cm^2 for each well. Samples of 0.5 to 1.5 ml were taken in intervals from 30 to 45 min and analyzed for reduction of resazurin as described in chapter 3.6.5. The top figure 5.1 shows the dynamic of resazurin reduction as calculated from equation 3.7 and the bottom figure 5.1 the dynamic as calculated from equation 3.10. To distinguish the individual experiments the total numbers of cells used are given. See table 5.1 for a reference concerning the cell number with respect to growth area. Filled circles in figure 5.1 correspond to experiments with just 5 ml starting volume for resazurin assay. Due to mixing with residual PBS++ from the washing step it is assumed that absorbances for resazurin species at $t = 0$ are significantly lower than those from batch medium, leading to a different value p for initial resorufin percentage (see equation 3.7). As initial absorbances from culture vessels were not measured but set to equal those from batch medium, calculation of accumulated resorufin therefore may result in a decline in mass of resorufin at first as it is the case for some experiments.

From figure 5.1, time points were selected for linear regression, whereby time points were excluded that belong to samples with difficulties during sampling or following analysis, as well as time points when reduction apparently converges saturation. Table 5.1 summarizes the results of linear regression.

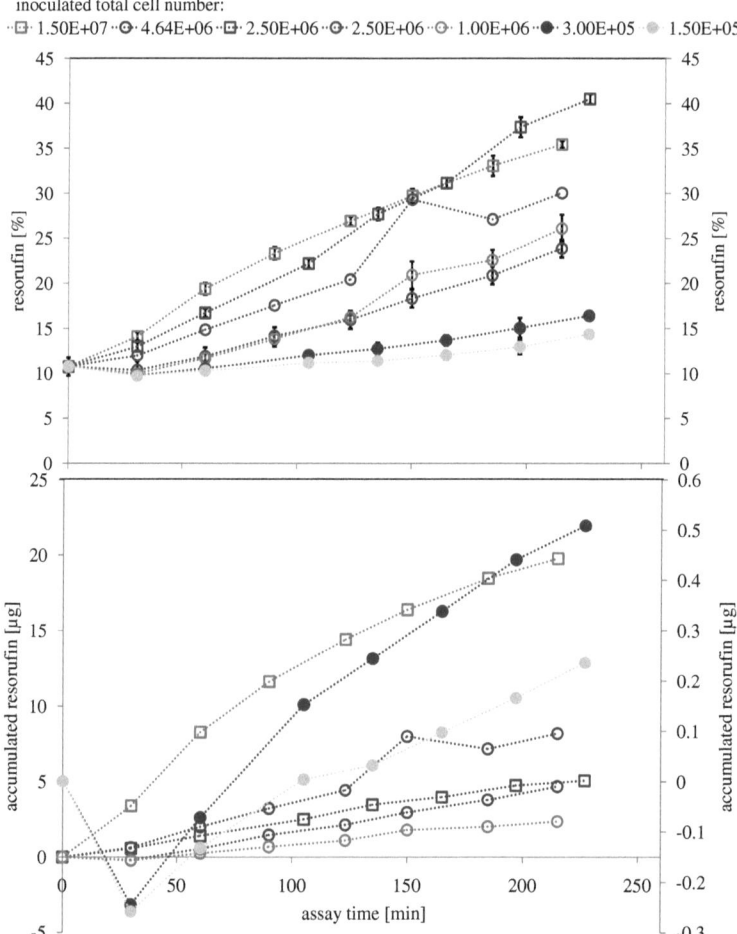

Figure 5.1: Dynamic of resazurin reduction for *CHO-K1*. Circles represent experiments where cell density was around $4 \cdot 10^4$ cells/cm^2, squares represent experiments with inoculated cell density around $8 \cdot 10^4$ cells/cm^2 or higher. Experiments were repeated with n as the numbers of repetitions varying from 1 to 6 as indicated in table 5.1. Top: Percentage of resorufin. Bottom: Accumulated resorufin, filled circles belong to vertical axis on the right side.

5. Analysis of CHO-K1 by resazurin assay and carbon content determination.

Table 5.1: Values for experimental evaluation of rate of reduction for *CHO-K1*.

inoculated cell number	cell density [cells/cm^2] (culture flask area)	n	time points for linear regression	slope [µg/min]	coefficient of determination (R^2)
1.5 · 10^7	8.6 · 10^4 (175)	2	30; 60; 90	0.137	0.988
4.64 · 10^6	6.2 · 10^4 (75)	1	30; 60; 90; 123	0.04	0.999
2.5 · 10^6	2.6 · 10^5 (9.6)	3	30; 60; 105; 135; 165	0.025	0.995
2.5 · 10^6	3.3 · 10^4 (75)	3	60; 90; 123; 150; 185; 215	0.026	0.997
1.0 · 10^6	4.0 · 10^4 (25)	3	60; 90; 123; 185	0.014	0.999
3.0 · 10^5	3.1 · 10^4 (9.6)	6	60; 135; 165; 197	0.003	0.994
1.5 · 10^5	1.6 · 10^4 (9.6)	6	135; 165; 197; 227	0.002	0.999

For all experiments, cell free medium containing resazurin was included as a negative control. Herein, spontaneous reduction of resazurin was not observed.

There is a strong coherence of the rate of reduction of resazurin as expressed in the slope of the reduction dynamic and the inoculated cell number. Figure 5.2 shows a linear relationship for a wide range of cell numbers. With a coefficient of determinination R^2 of 0.997, following linear regression the cell number can be evaluated according to equation 5.1.

Obviously, the growth phase does not influence the metabolic status regarding resazurin reduction, as for 2.5 · 10^6 cells inoculated with a high cell density (stationary growth phase) the same slope was observed as for 2.5 · 10^6 cells inoculated with a low density (exponential growth phase), see table 5.1. Moreover, the slope for 1.5 · 10^7 cells (stationary growing) also is in alignment, see figure 5.2.

$$n_{cell} = \frac{\text{slope (in µg/min)} - 0.00182\,\text{µg/min}}{0.009\,\text{µg/min}} * 10^6 \tag{5.1}$$

As with increasing accumulating mass of resorufin the percentage of reduced resazurin approaches saturation, for later experiments it is strongly recommended to evaluate the reduction dynamic rather than an endpoint value. Out of the first measurements, a time range should be estimated in which reduction is nearly linear and the slope should be determined for that range as shown above. Then, equation 5.1 gives a good hint of the cell number n_{cell} of underlying experiments.

For endpoint assays, however, the slope can be estimated by dividing the difference in resorufin content for the start and end of assay by the assay time. The determined cell number as evaluated by equation 5.1 is then a lower bound for cell number taking part in the assay, at

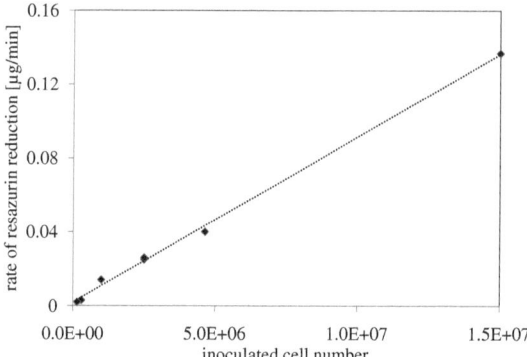

Figure 5.2: Slope determined from figure 5.1 plotted against inoculated cell number.

least.

5.1.2 Modeling resazurin reduction

To gain a more elaborate insight into resazurin reduction, data from chapter 5.1.1 was evaluated to design a mathematical model of resazurin reduction in collaboration with G. Gelbert, TU Berlin. The model describes the percentage of resorufin over time depending on the underlying cell number for resazurin assays. Model assumptions were formulated as follows:

- The dynamic of resazurin reduction to resorufin was set to follow a kinetic with a power law influence of resazurin concentration and a linear influence of cell number. Linear reduction within the tested frame was assured experimentally, see figure 5.2. However, for adherently growing cells, metabolic rates are not only determined by cell number but also by cell density on a growth surface. This influence will not be modeled.

- Further reduction of resorufin to hydroresorufin was assumed to be very slow for low concentrations of resorufin, therefore, reduction was set equal to zero. Furthermore, hydroresorufin, which most likely occurs at very low concentrations in batch medium at $t = t_0$, but is not determined experimentally, is assumed to equal zero at assay start.

- Reduction of resazurin is assigned to enzymatical catalysis by cells solely. This assumption was validated in experiments where cell and ceramic free resazurin medium was analyzed in a time frame of several hours without significant increase in resorufin (data not shown).

- Cell number was assumed to be constant during assay time, as the population time of *CHO-K1* is around 20 h.

The deduced model equations read as follows:

$$\frac{dm_R}{dt} = -\mu_{max} * x_R^K * n_{cell} \tag{5.2}$$

$$\frac{dm_F}{dt} = \mu_{max} * x_R^K * n_{cell} \tag{5.3}$$

where m_R and m_F are the total mass (in µg) of resazurin and resorufin, respectively, $x_R = \frac{c_R}{c_R^*}$ is a normalized concentration of resazurin with c_R being the concentration of resazurin and $c_R^* = 1\,\text{µg/ml}$. c_R is calculated by $c_R = \frac{m_R}{V}$ with V being the experimental volume (in ml). n_{cell} is the number of cells and μ_{max} and K serve as model parameters. Due to sampling, V is reduced during the experiment, which was taken into account for the model. Percentage of resorufin was calculated by $p_F = 100 * \frac{c_F}{c_{sum}}$, whereby c_F is the concentration of resorufin and c_{sum} the total concentration of all reaction species resazurin and resorufin which is $5\,\text{µg/ml}$. Model parameters were identified from 4 out of 7 experiments described in chapter 5.1.1 using Matlab's "fminsearch" optimization with a least squares error. As some experiments showed an initial drop of percentage of resorufin (see figure 5.1) probably due to mixing with PBS++ as discussed in chapter 5.1.1, simulations for parameter identification and later also validation were performed from $t = 30\,\text{min}$ which was the time point of first measurement of medium out of each culture vessel. Parameters obtained read as follows:

$$\mu_{max} = 2.5 \cdot 10^{-10}\,\text{µg}\,\text{min}^{-1}\,\text{cell}^{-1} \tag{5.4}$$

$$K = 2.7 \tag{5.5}$$

Figure 5.3 shows the experiments used for parameter identification. The quality of the model was validated by simulating three experiments from chapter 5.1.1 not used for parameter identification, see figure 5.4.

Acceptable results were obtained for modeling resazurin reduction, see figure 5.4 and 5.3. However, measurement data from the experiment with a very high cell number of $1.5 \cdot 10^7$ showed a slowdown in resazurin reduction during the assay which might be attributed to a decrease in metabolic rates. As reduction of medium volume during sampling here was performed faster than for all the other experiments, lactate accumulation and/or lack of glucose in the culture vessel could have occurred, resulting in a decrease in cell metabolism during the course of the experiment.

5.1.3 Adaptation of the resazurin reduction model to describe bioreactor performance

For resazurin assays performed inside the reactor modules, a simple model was designed that returns the percentage of reduced resazurin depending on the underlying growth performance of cells in the reactor. Therefore, the total volume of the reactor was divided into three main compartments being A the medium container, B the foam containing magazine where the cell catalyzed reduction takes place, and C a small volume behind the magazine from where

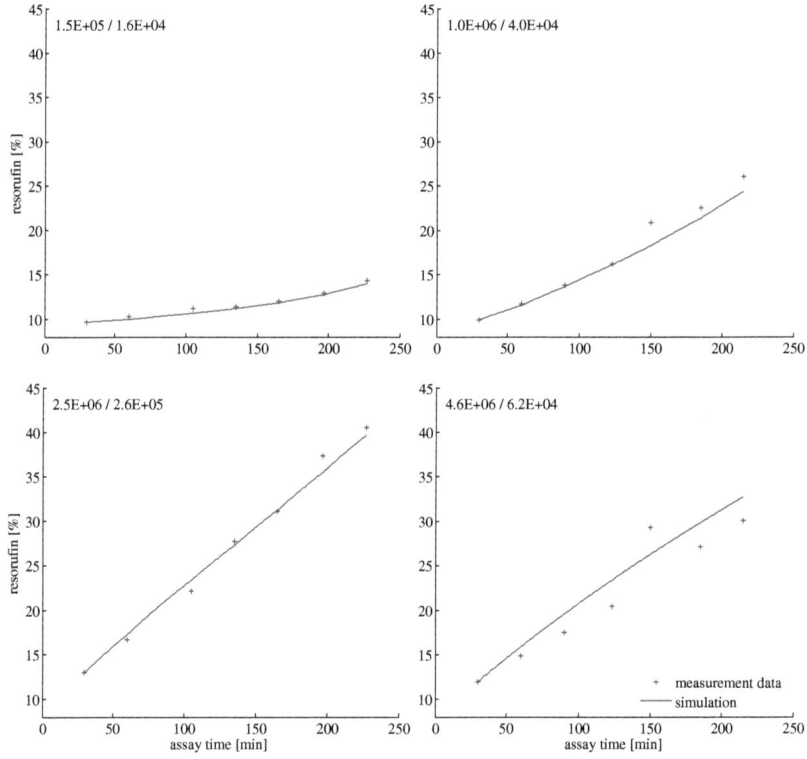

Figure 5.3: Model fit for resazurin reduction by *CHO-K1*. Figures show experimental data and simulation results for experiments used for parameter identification, cell number and cell density in cells /cm² is indicated on top of each panel.

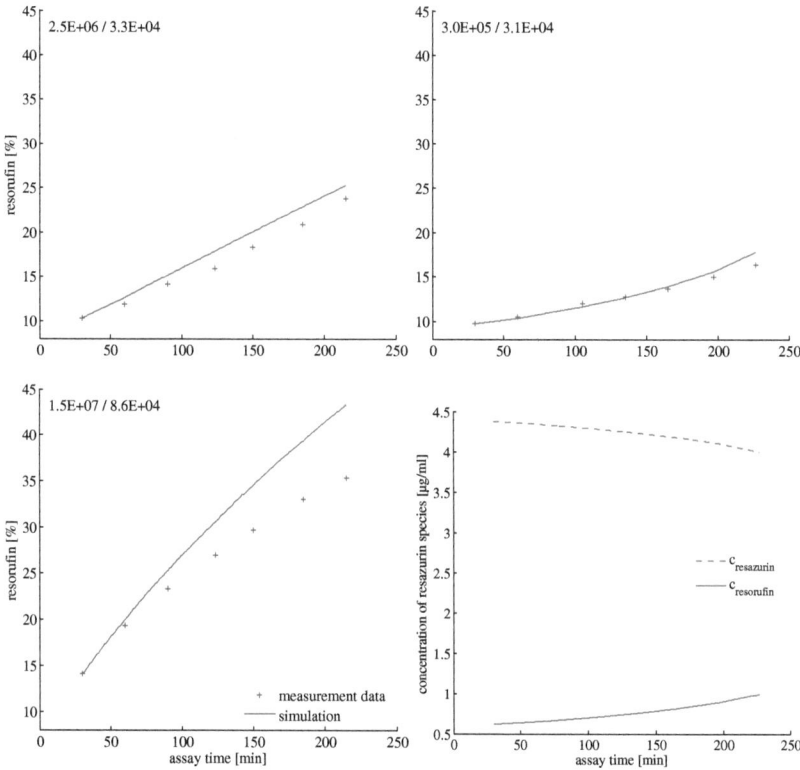

Figure 5.4: Model validation for resazurin reduction by *CHO-K1*. Figures show experiments not used for parameter identification. Cell number and cell density in cells/cm² is indicated on top of each panel. Bottom right figure shows the concentration development for resazurin and resorufin for the panel above.

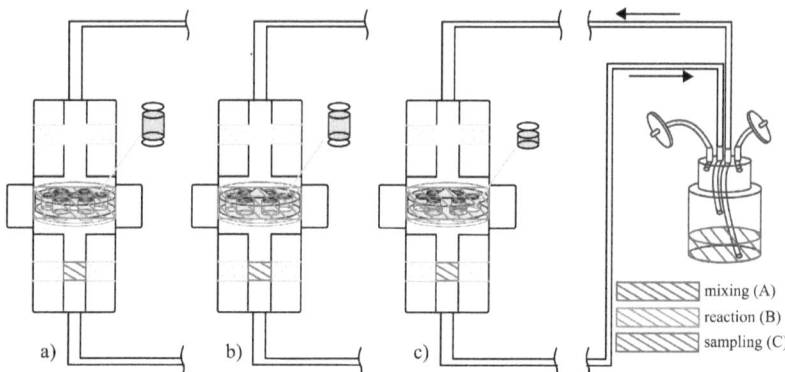

Figure 5.5: Scheme for division of reactor volume into compartments for flow modeling. Three different volumes for A, B and C, respectively, were calculated depending on magazine geometry and ceramics' height. a) Geometry for 7 ceramics with 10 mm height in plain magazine, b) geometry for 6 ceramics with 10 mm height in conical magazine, c) geometry for 6 ceramics with 5 mm height in conical magazine.

sampling is performed (see figure 5.5). Model assumptions read as follows:

- Flow inside modules and periphery is assumed to be laminar without backmixing or diffusion of reaction species. Modules are specified as ideal plug flow reactors.

- Dotted areas in figure 5.5 do not take part in reaction or flow although filled with medium; diffusion from those areas is assumed to equal zero.

- Cells are assumed to be equally distributed inside compartment B. A medium volume entering compartment B is immediately mixed ideally and supposed to leave B directly.

- Ideal mixing is assumed for compartment A.

- Sampling is performed from compartment C, reducing the medium volume in the tube going to compartment A. Hereby, backmixing in C is assumed to be zero and the tube to A fills with air from the medium container.

- Reduction of resazurin is only attributed to enzymatic activity of cells. This assumption is valid, as no reduction of resazurin was observed during a resazurin assay inside a reactor module with 6 cell free ceramics over a period of 3 h.

Moreover, model assumptions as stated in chapter 5.1.2 apply.

Volumes were determined experimentally and by help of mechanical drawings of modules (see Appendix), confer to table 5.2. Total medium volume inside the reactor and periphery at $t = t_0$ was 23 ml for module type a and 40 ml for module types b and c in figure 5.5, flow velocity was 1.4 ml min^{-1} for all simulated experiments. For simplification, the reaction was assumed not to start before the reactor volume was totally filled with medium (see chapter

Table 5.2: Conditions for modeling resazurin reduction inside bioreactors. Lag times $T_{A,0}$, $T_{B,0}$ and $T_{C,0}$ were calculated for a flow of 1.4 ml/min and starting medium volume of 40 ml.

module type as in figure 5.5	volume A ($t = t_0$) [ml]	volume B [ml]	$T_{A,0}$ [min]	$T_{B,0}$ [min]	$T_{C,0}$ [min]
a)	5.8	4.9	8.7	5.6	4.6
b)	16.5	4.2	8.7	5	4.1
c)	17.1	2.1	9.8	5	4.1

3.6.5). During dynamic resazurin assay, the total volume is reduced due to sampling, which had to be taken into consideration for the model. Therefore, the model is separated into two parts. One describing operation between two samples (I) and one describing the process after sample removal (II). In part I medium circulates all over three compartments A, B, and C and all tubes are thoroughly filled with medium. Resorufin concentration profiles are described by model equations 5.6 to 5.10. Following sampling, the tube going to compartment A is partly filled with air, which was sucked in during the sampling process. Therefore, no medium passes into A for a time period of $t_{sample} = \frac{\text{sample volume}}{\text{flow velocity}}$. Hence, in part II equations 5.9 to 5.10 and equations 5.11 to 5.13 are solved for a time period t_{sample}.

Following part II, all compartments and tubes are filled with medium again but as a liquid volume has been removed during sampling, delay times of part I do not apply for equations 5.6 an 5.7. Therefore, at first equations of part I are solved with a revised delay time $T'_{B,0}$ for a duration t' until the back drawn medium entirely leaves the tube to compartment A (part I'). Hereby $T'_{B,0}$ is calculated as $T_{B,0} + t_{sample}$, whereby $T_{B,0}$ is the lag time needed for a volume element of B to approach to compartment A, see table 5.2. t' is calculated as $T_{B,0} - T_{C,0} - t_{sample}$ with $T_{C,0}$ being the lag time needed for a volume element of B to approach to compartment C. Then, delay times of part I apply again and model solving continues with equations from part I as above.

I:

$$\frac{dm_{A,R}}{dt} = \dot{V} * (c_{B,R,0} - c_{A,R}) \tag{5.6}$$

$$\frac{dm_{A,F}}{dt} = \dot{V} * (c_{B,F,0} - c_{A,F}) \tag{5.7}$$

$$\frac{dV_A}{dt} = 0 \tag{5.8}$$

$$\frac{dm_{B,R}}{dt} = \dot{V} * (c_{A,R,0} - c_{B,R}) - \mu_{max} * x^K_{B,R} * n_{cell} \tag{5.9}$$

$$\frac{dm_{B,F}}{dt} = \dot{V} * (c_{A,F,0} - c_{B,F}) + \mu_{max} * x^K_{B,R} * n_{cell} \tag{5.10}$$

II:

$$\frac{dm_{A,R}}{dt} = -\dot{V} * c_{A,R} \tag{5.11}$$

$$\frac{dm_{A,F}}{dt} = -\dot{V} * c_{A,F} \tag{5.12}$$

$$\frac{dV_A}{dt} = -\dot{V} \tag{5.13}$$

Variables are calculated as

$$c_{y,z} = \frac{m_{y,z}}{V_y} \tag{5.14}$$

$$c_{y,z,0} = \frac{m_{y,z,0}}{V_{y,0}}, \tag{5.15}$$

where $m_{y,z}$ and $c_{y,z}$ are mass and concentration of resazurin species z (resazurin and resorufin) in compartment y (A, B), respectively, $x_{B,R} = \frac{c_{B,R}}{c_{B,R}^*}$ is a normalized concentration with $c_{B,R}^* = 1\,\mu g/ml$. $\dot{V}$ is the volumetric flow during the assay. $c_{B,z,0}$ is the concentration of resazurin species z in compartment B at time $t - T_{B,0}$. Likewise, $c_{A,z,0}$ is the concentration of resazurin in compartment A at time $t - T_{A,0}$ as well as $V_{A,0}$ is the volume in compartment A at time $t - T_{A,0}$, where $T_{A,0}$ is the lag time needed for a volume element of A to approach to compartment B. V_y is the total volume of medium in compartments A and B, respectively. n_{cell} is the number of cells taking part in the reaction and μ_{max} and K are the model parameters as identified in chapter 5.1.2.

Concentrations of reaction species in compartment C were calculated to be those of compartment B at time $t - T_{C,0}$.

The model is solved with "dde23" in Matlab. For validation, $3.15 \cdot 10^6$ cells/foam were statically inoculated on dry standard foams as in chapter 3.3.1 and allowed to adhere for 2 h. Six foams were then assembled aseptically into reactor module type b) as in figure 5.5, washed with PBS++ and subjected to a resazurin assay as in chapter 3.6.5. The experimental outcome and simulation results are depicted in figure 5.6. Simulation and experimental data were in adequately good agreement, so for the experiments described later in this work this model can be considered a good first evaluation of the dynamics of resazurin reduction and therefore allows an estimation of the cell number.

5.2 Carbon content of cells cultivated on ceramics

5.2.1 Carbon content of pure cells

In order to determine a correlation between cell count and measurable carbon content, cells were centrifuged, counted by a hemocytometer and aliquots of cells in PBS were subjected to carrier hot gas extraction (CHGE) as in chapter 3.6.6. Total carbon and sulfur content are depicted in figure 5.7, whereby cell free samples were below detection limits and therefore excluded

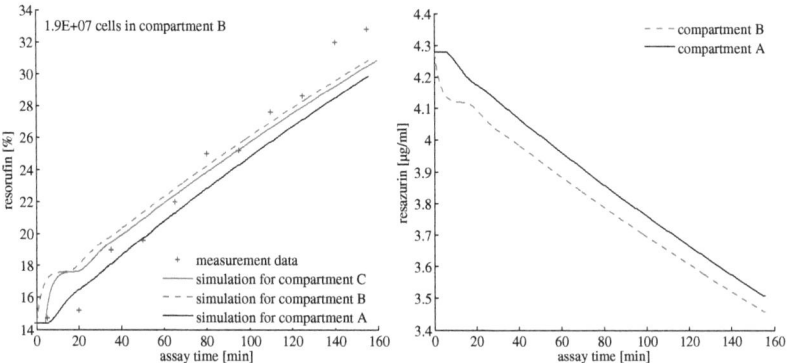

Figure 5.6: Model validation for reduction of resazurin inside the bioreactor. Solid lines represent simulation data, + is for experimental measurement data. Left: percentage of reduced resazurin (resorufin) in reactor compartments. Right: Simulation of concentration development for resazurin in compartments A und B.

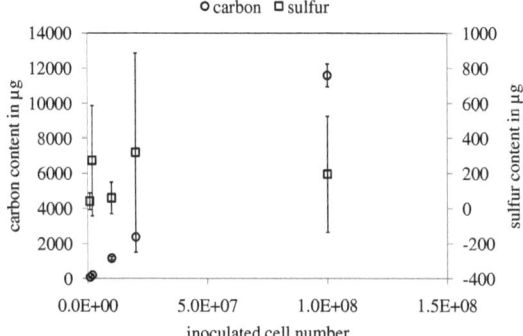

Figure 5.7: Carbon and sulfur content of *CHO-K1* in PBS. The number of repetitions n ranges from 3 to 6.

for illustration. Whereas carbon content correlated very good with cell number (regression coefficient 0.999), analyzed sulfur content did not correlate with cell number at all, see sulfur measurements and standard deviations in figure 5.7. From carbon measurements in figure 5.7, the following linear regression for the cell number n_{cell} can be estimated:

$$n_{cell} = \frac{m_c}{0.000116\,\mu g}, \quad (5.16)$$

where m_c is analyzed carbon content in µg.

5.2.2 Carbon content of cells on foams

To determine the influence of ceramics and medium composites on carbon analysis, ceramic cylinders of 10 mm height, 10 mm diameter and with one blind hole from above with 6 mm diameter

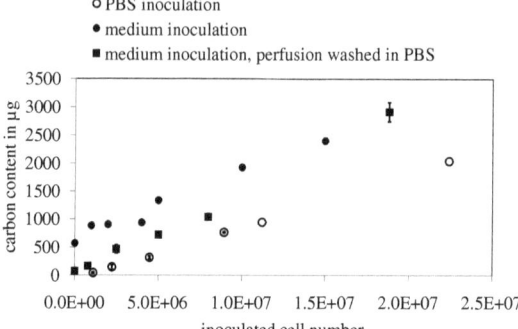

Figure 5.8: Carbon content of *CHO-K1* on ceramics. Cells were inoculated in PBS or medium as stated in legend and dried prior to CHGE. For medium inoculated cells, ceramics were washed in PBS before drying, see text for further explanation. n ranges from 1 to 3.

and 7 mm depth were prepared as in chapter 3.1 with additional teflon sealing for the ceramics' bottoms. Ceramics were dried overnight at 50°C in a laboratory oven prior to static inoculation of cells in PBS or culture medium.

For cells inoculated in PBS, foams with removed teflon tape were dried over night at 50°C in a laboratory oven directly after inoculation.

Cells inoculated in medium were allowed to settle for 135 min, then the teflon tape was removed and ceramics were washed in PBS and dried overnight at 50°C. To simulate cultivation of cells on ceramics under perfusion, some ceramics were analyzed for resazurin reduction 120 min after inoculation in medium. Single ceramics hereby were assembled sterile into the small block reactors as in chapter 3.2.3 and perfused with 6 ml culture medium containing 22 µmol/l resazurin for 165 min. For this, the teflon tape sealing at the ceramics' bottom was removed. Total medium was collected and analyzed for resazurin reduction as in chapter 3.6.5 and foams inside the reactor modules were rinsed with 30 ml PBS for 20 min. Ceramics were disassembled, the teflon tape was removed and foams were dried overnight at 50°C.

Ceramics were stored at -80°C prior to CHGE as in chapter 3.6.6 and carbon content was calculated according to equation 3.13.

Whereas carbon content of cell free ceramics incubated in PBS was below detection limit, cell free ceramics incubated in medium showed a carbon content of about 500 µg per foam, although ceramics were thoroughly washed in PBS prior to CHGE, see filled circles in figure 5.8. This number was reduced to only 70 µg per foam if foams were perfused with PBS in small block reactors (filled squares in figure 5.8). Therefore, the elimination of residual medium inside ceramics' pores is a crucial step for CHGE and was performed for all subsequent experiments utilizing reactor modules by perfusion washing in PBS for 30 min.

Ceramics with cells inoculated in PBS without further cultivation showed a linear correlation between carbon content and inoculated cell number. Similar results were obtained for ceramics with cells inoculated in medium and cultivated for another 285 min (partly during resazurin

Table 5.3: Linear regression for carbon analysis as in figure 5.8. n_{cell}, cell number; * marks linear regression as for the experiment in the above row but with excluded value for $1.88 \cdot 10^6$ cells/foam, see text for explanation.

method	carbon content following linear regression [µg]	coefficient of determination (R^2)
inoculation in PBS, no further cultivation; n=1-2	0.000088*n_{cell}	0.993
inoculation in medium, washing in PBS, cultivation for 135 min; n=1	0.000119*n_{cell}+647	0.972
inoculation in medium, cultivation for 135 min, resazurin assay for 165 min, perfusion washing in PBS; n=2-3	0.00015*n_{cell}+21	0.991
inoculation in medium, cultivation for 135 min, resazurin assay for 165 min, perfusion washing in PBS*; n=2-3	0.000122*n_{cell}+96	0.989

assay) prior to perfusion washing with PBS and subsequent CHGE. Following resazurin assays, ceramics showed a spectrum of colors ranging from blue to pink, depending on reduction of resazurin and inoculated cell number. Nevertheless, this variation in color did not influence the outcome of carbon analysis, see filled squares in figure 5.8. However, ceramics with cells inoculated in medium and washed in PBS without perfusion after 135 min of cultivation did not show such a good correlation concerning carbon content and cell number (filled circles in figure 5.8), probably due to poor medium elimination as ceramics solely were flushed in PBS for washing. Table 5.3 gives an overview on calibration curves deduced from figure 5.8.

Nevertheless, for those foams with experimental performance closest to the following actual experiments (filled squares in figure 5.8) foams with the highest inoculated cell number of $1.88 \cdot 10^6$ cells/foam seemed to match more the trend of foams not so thoroughly washed (filled circles). This most probably is attributed to the inoculation method, where a highly concentrated cell suspension is dropped onto the foam, leading to cell blockage with therefore poor washing in PBS and hereby unwanted measurement of medium species in CHGE. Therefore, this value was excluded for linear regression (see last row in table 5.3) leading to equation 5.17 for the calculation of cell number n_{cell} for the following experiments.

$$n_{cell} = \frac{m_c - 96\,\mu g}{0.000122\,\mu g}, \quad (5.17)$$

where m_c is analyzed carbon content in µg.

Equation 5.17 clearly only applies if cell distribution on and in ceramics is homogeneous without cell blockage, allowing efficient PBS-washing before carbon analysis in CHGE. Moreover, abundance of matrix proteins as described in chapter 7.1 and chapter 8 will influence cell number estimation derived from equation 5.17 significantly due to their high carbon content, which will be assigned to a high cell number mistakenly.

6 Influence of mode of inoculation on cellular growth and distribution

As discussed in chapter 2.1.3 initial cellularity of a porous scaffold strongly influences further cell growth inside the scaffold. To obtain a high number of cells homogeneously distributed all over the ceramic volume different approaches of inoculation were evaluated. Besides experiments described below, I also employed two gel inoculation techniques hoping for strong cell penetration of the ceramic due to impeded cell flush out. Whereas [Li et al., 2008] and [Blan and Birla, 2008] found promising results, in the system at hand neither gelation of a 1% sodium alginate (Fluka, 71238)/cell suspension in ceramics with 0.2 mM calcium chloride (Merck, 1.02381) nor application of different concentrations of Matrigel (BD Biosciences, 356234) in cell suspensions on ceramics resulted in even cell distribution. Moreover, cells were found to die inside the viscous matrix, even though flow channels were drilled into the constructs to improve nutrient access. Cell number and distribution inside the ceramics was found to be the poorest of all realized experiments, therefore a more detailed experimental description of gel inoculation was neglected for the current chapter.

6.1 Static inoculation

6.1.1 Static inoculation on foams in culture plates

To prove the ability of cells to proliferate on alumina, *CHO-K1* were statically inoculated as in chapter 3.3.1 with $2.7 \cdot 10^5$ and $1.6 \cdot 10^5$ cells/foam, respectively, and statically cultivated (chapter 3.4.2) for five to eight days. Ceramic foams were transferred to new 12-well culture plates before resazurin assay to avoid analysis of cells attached to the bottom of the well. Resazurin assays (see chapter 3.6.5) were performed for 4 h at selected days with results depicted in figure 6.1.

Metabolic reduction of resazurin clearly increased during cultivation time for $2.7 \cdot 10^5$ and $1.6 \cdot 10^5$ cells/foam inoculation densities, respectively. Moreover, with equation 5.1 on page 39, mean cell number per foam was estimated to be $3.9 \cdot 10^6$ for inoculation with $2.7 \cdot 10^5$ cells and cultivation for 8 days, and to be $1.3 \cdot 10^6$ for inoculation with $1.6 \cdot 10^5$ cells and cultivation for 5 days. Therefore, cell number apparently multiplied by factor 14 and 8, respectively, and proliferation can be assumed. The increase in cell number, however, is significantly below the theoretical limit assuming a doubling time of 20 h and absence of contact inhibition. Likewise, dyeing of cells on the ceramic surface and microscopic evaluation as in chapter 3.6.2 displayed

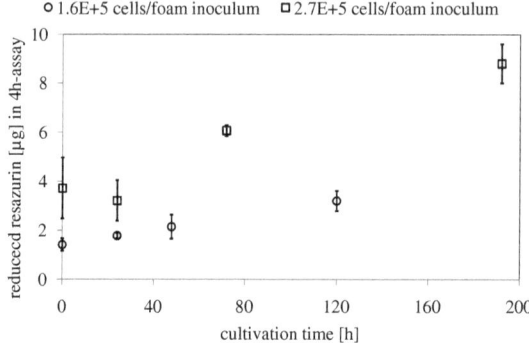

Figure 6.1: Mass of reduced resazurin (n=3) after 4 h resazurin assays for two inoculation densities following static inoculation and static cultivation of cells on ceramic foams in culture plates.

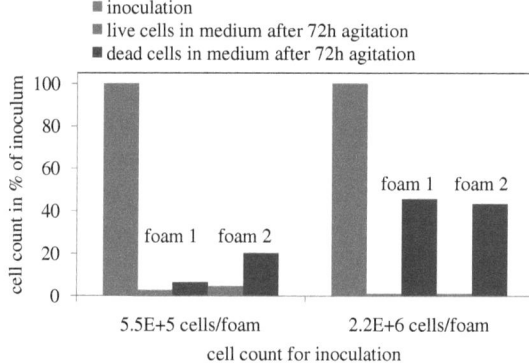

Figure 6.2: Cell leakage for cells statically inoculated on ceramic foams in culture plates and cultivated with agitation for 72 h. Shown is cell count in medium for two foams for each inoculation density.

poor cellular growth on the lower side of the foams and cellular ingrowth only to approximately 200 µm. In another experiment, foams were subjected to agitation cultivation as in chapter 3.4.3 210 min after static inoculation with $5.5 \cdot 10^5$ and $2.2 \cdot 10^6$ cells/foam. After one, three and five days of cultivation, no cellular growth deeper than 200 µm was observed for all foams. Moreover, cells attached to the foam appeared less vital and huge cell numbers were observed in the surrounding medium (as counted by help of a hemocytometer), see figure 6.2.

6.1.2 Static inoculation into the tubular reactor

For static inoculation into tubular reactors (see figure 3.3), 4 ceramic foams were assembled into the reactor in such a way that the first foam's upside was located next to the upper inoculation port. Following preparation as in 3.2.1 and vertical alignment of the module in the climatic chamber, $1 \cdot 10^7$ *CHO-K1* cells were inoculated directly onto the first foam and allowed to

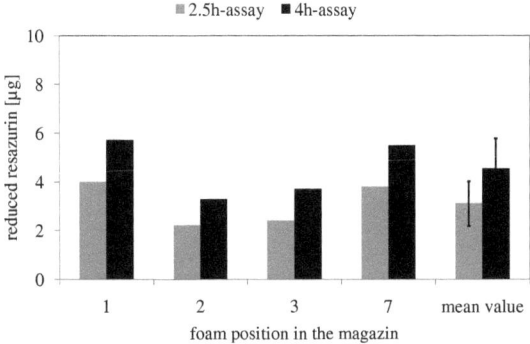

Figure 6.3: Mass of reduced resazurin in a 2.5 and 4 h resazurin assay for static inoculation with $1 \cdot 10^6$ cells/foam in a revolver reactor after 31 days of perfusion cultivation.

settle for 4 h before perfsuion was imprinted from above at $0.08 \, \text{ml min}^{-1} \, \text{cm}^{-2}$ for 20 d. The module was then disassembled and single foams were subjected to microscopic evaluation as in chapter 3.6.2. While disassembling, thick clumps of cells and cell debris were found between the sealing rings and the module shell, probably remaining cells from inoculation. Nevertheless, cells were found to be very dense on top of the first foam, and much lesser dense on all other top and bottom sides of the foams with very inhomogeneous cell distribution. No cells were found in the foam's insides.

6.1.3 Static inoculation into revolver reactors

For static inoculation into revolver reactors with plain magazines (see figure 3.6a), seven ceramic foams were inoculated at once with $7 \cdot 10^6$ *CHO-K1* cells as in chapter 3.3.1 and cultivated for 31 d under perfusion conditions with $0.14 \, \text{ml min}^{-1} \, \text{cm}^{-2}$ as described in chapter 3.4.1. The reactor module was then disassembled and three foams from the magazine's rim (positions 1-3) as well as the foam in the middle of the magazine (position 7) were subjected to a resazurin assay for 2.5 and 4 h, see figure 6.3. Obviously, inhomogeneous cell cultivation inside the reactor module led to very different cell numbers on single foams, as metabolic reduction of resazurin fluctuated strongly between foams.

Whereas all foams' upsides and even the foam holding magazine's upside were covered with a thick biofilm (which turned out to contain dead cells in a mucous matrix), total cell number seemed to be rather low - by equation 5.1 on page 39 a mean cell number of $2.0 \cdot 10^6$ cells/foam was calculated from overall resazurin reduction in the 2.5 h assay. Compared to ceramics statically inoculated with only $2.7 \cdot 10^5$ cells/foam and cultivated under static conditions, cell growth therefore seemed to be very slow (see figure 6.1). Moreover, FDA/EB staining as described in chapter 3.6.1 revealed a dense layer of vital cells on the foams' upside, very few and less vital cells at the bottom, see figure 6.4, and no cells inside the ceramic (for expressive colored images please visit http://opus.kobv.de/tuberlin/volltexte/2010/2756/).

Nevertheless, to evaluate the influence of flow velocity on cellular growth, in a further experi-

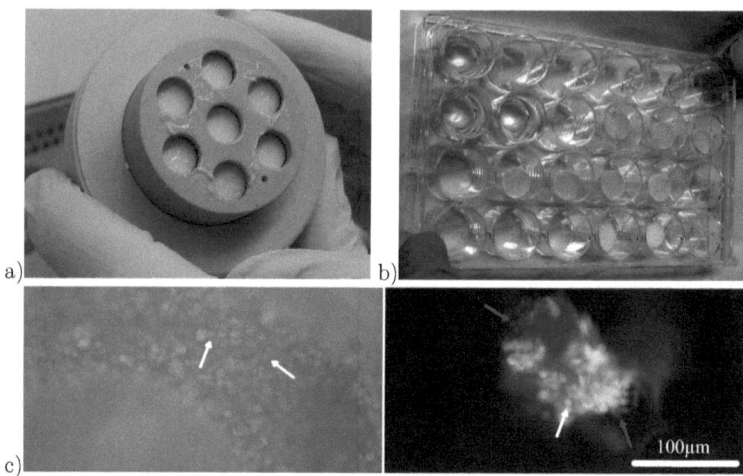

Figure 6.4: Optical analysis of ceramic foams statically inoculated into a revolver reactor and followed by 31 d of perfusion cultivation. a) Biofilm on foam holding magazin. b) Biofilm on the upside of all 7 single foams after disassembling ceramics from the magazine, ceramics in the culture plate's last row are a cell free control. c) FDA/EB staining for cells on one representative foam's upside (left) and on the bottom side of that foam (right). Vital cells fluoresce green, dead cells red, white arrows point to single vital cells, red arrows point to single dead cells.

ment $1 \cdot 10^7$ cells/module were inoculated into five reactor modules carrying 7 foams each and allowed to settle for 3 h before perfusion took place. Perfusion was then carried out at 0.14, 0.29, 0.33, 0.44 and 0.61 ml min^{-1} cm^{-2}, respectively, for 13 days. Foams were analyzed in a vitality assay (chapter 3.6.1) and glucose consumption and lactate formation were monitored (chapter 3.6.4). A flow velocity of 0.29 ml min^{-1} cm^{-2} hereby supported cellular growth best as there was the highest glucose comsumption and lactate formation over a culture period of 13 d (see figure 6.5). The same results were obtained for 0.33 ml min^{-1} cm^{-2} (data not shown). At 0.44 ml min^{-1} cm^{-2} an equal consumption was found for glucose and production of lactate as for 0.29 or 0.33 ml min^{-1} cm^{-2} (data not shown) and foams also yielded good growth, but FDA/EB staining clearly revealed that cell vitality (i. e. percentage of vital cells over total cell number) did not approach that of the lower flow velocity. At a flow rate of 0.61 ml min^{-1} cm^{-2} cell metabolism was slightly reduced compared to 0.29 ml min^{-1} cm^{-2}, and total cell density was lower. Moreover, dyed cross sections revealed a relation between flow velocity and cellular ingrowth into the foam. For 0.14, 0.44 and 0.61 ml min^{-1} cm^{-2} almost no vital cells were found in 5 mm depth, whereas cell colonies were observed for 0.29 and 0.33 ml min^{-1} cm^{-2} (see figure 6.6).

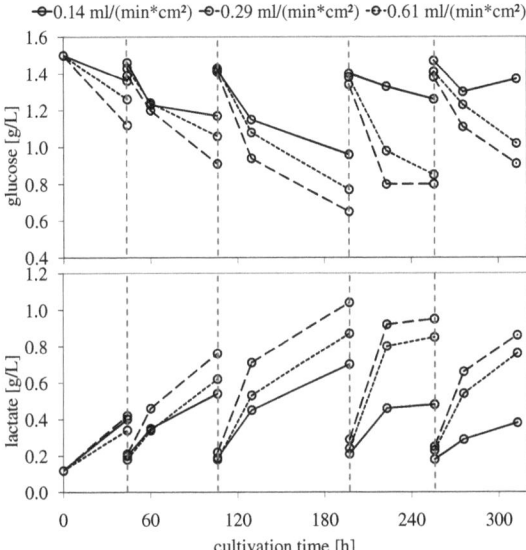

Figure 6.5: Glucose consumption and lactate formation for *CHO-K1* statically inoculated into reactor modules and cultivated with different perfusion velocities for 13 d ($1 \cdot 10^7$ cells/reactor module, 7 foams/module). Dashed horizontal lines indicate medium replacement.

6.1.4 Static inoculation in ceramics with flow channels

Obviously, the filter-like ceramic shape hinders cells to penetrate the foam's inside during static inoculation. Therefore, in another experiment, I introduced a ceramic with blind holes from both face areas to provide a larger, accessible surface. As these experiments were performed very early in the project's progress, *CHO-K1* adapted to growth in medium containing 1% FBS (see chapter 3.5 and Appendix) were used, which are very sensitive to flow velocities higher than $0.12\,\text{ml}\,\text{min}^{-1}\,\text{cm}^{-2}$. Thus, this section should be regarded as one individual unit and not be compared to experimental outcomes of all other chapters.

Revolver modules with plain magazines were prepared with 7 foams per magazine each, whereby ceramics were modified as follows: Prior to the last torching step in chapter 3.1, four blind holes with 7 mm depth and 0.2 mm diameter were drilled into the ceramic cylinders from the upside of the foam and four blind holes were drilled from the bottom. Blind holes were aligned in a square with distance from the foam's middle 1.5 mm for the upside and 3 mm from the bottom. By that, holes did not touch each other and flow was forced to pass the foams' 3D structure. Module assembly otherwise was performed as described in chapter 3.2.2.

For static inoculation, $1 \cdot 10^6$ *CHO-K1* cells adapted to growth in medium containing 1% FBS were inoculated per foam from above and cells were allowed to settle before perfusion took place with flow velocities 0.03, 0.08 and $0.12\,\text{ml}\,\text{min}^{-1}\,\text{cm}^{-2}$, respectively. Modules were subjected to perfusion cultivation for 1 and 7 days, then, foams were separately subjected to resazurin assays in culture wells as in chapter 3.6.5 and medium was analyzed for glucose consumption

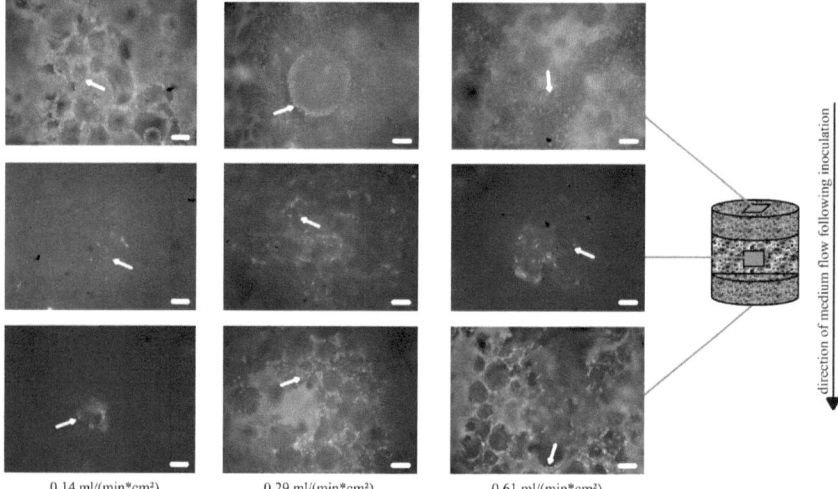

Figure 6.6: FDA/EB staining of *CHO-K1* cells statically inoculated into reactor modules and cultivated with different perfusion velocities for 13 days ($1 \cdot 10^7$ cells/module inoculum, 7 foams/module). Green fluorescence corresponds to vital cells, red fluorescence to dead cells, bars represent 100 μm, arrows point to single cells. Each column represents foams of a specific perfusion velocity during cultivation as indicated underneath. For each flow velocity representative pictures are shown for the foam's face (top row), a cross section in about 4 mm depth (middle row) and the end area (bottom row).

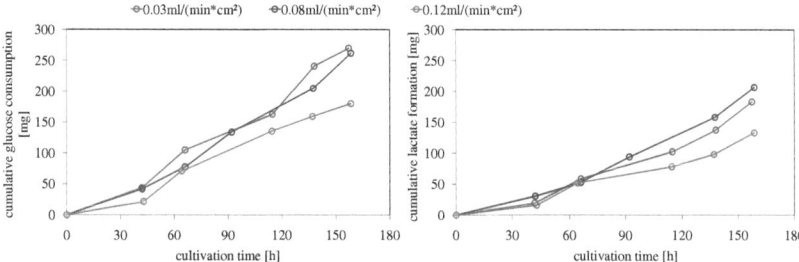

Figure 6.7: Cumulative glucose consumption (left) and lactate formation (right) for CHO-K1 adapted to growth in medium with reduced serum content (1% FBS) statically inoculated into revolver reactors (7 · 10^6 cells/module, 7 foams/module) equipped with foams with blind holes (n=1). Cultivation was performed for 7 days at three flow velocities as indicated. Medium replacement was performed twice for 0.03 and 0.08 ml min^{-1} cm^{-2} on day 3 and day 5 and three times for 0.12 ml min^{-1} cm^{-2} on day 2, 4 and 6.

and lactate formation. Single foams were also subjected to vitality assays (FDA/EB staining). As a control, 1 · 10^6 cells/foam were statically inoculated (chapter 3.3.1) on foams modified as above and statically cultivated (chapter 3.4.2) for 1 and 7 days, respectively.

Regarding overall glucose consumption and lactate formation, there were only small differences for the three flow velocities, see figure 6.7. However, for 0.08 ml min^{-1} cm^{-2}, lactate formation occurred slightly faster than in the other modules. As for single foam analysis for 3 h following cultivation, glucose consumption and lactate formation differed strongly for foams cultivated for 1 or 7 days, see figure 6.8. However, there was almost no difference between glucose consumption between the flow velocities, lactate formation was slightly higher for 0.03 ml min^{-1} cm^{-2} and 7 days of cultivation. Reduction of resazurin in a 4 h assay again revealed an increase in metabolic activity after 7 days of cultivation, see figure 6.9. Here, 0.08 ml min^{-1} cm^{-2} showed the highest increase for the three flow velocities. Remarkably, both medium and resazurin assays showed the highest increase in metabolic activity for the control group.

However, vitality assays of cross sections revealed cellular ingrowth into the blind holes both from top and bottom for all foams after 1 day of cultivation. Following 7 days of inoculation, growth was observed in pores next to the blind holes for 0.08 ml min^{-1} cm^{-2} flow velocity, but not for any other foams, see figure 6.10. Moreover, cells were found in the blind holes from top in all foams after 7 d, but only for 0.08 ml min^{-1} cm^{-2} and the control group, vital cells were found in bottom blind holes. Consistent with analyses depicted in figures 6.7, 6.8 and 6.9, 0.08 ml min^{-1} cm^{-2} yielded best results regarding 1 week long cultivation. Again, foams of the control group showed comparable good results for vitality staining, as growth was observed inside bottom blind holes after 7 days of static cultivation.

Still, it has to be mentioned that medium and resazurin assays as performed in sections 6.1.1 to 6.1.4 do not properly evaluate cells growing inside the porous structure. Here, assays were performed statically, i. e. cells growing inside the matrix do not take part in metabolic processes, or, if they do, medium is not analyzed as it clearly cannot be washed out thoroughly. Therefore,

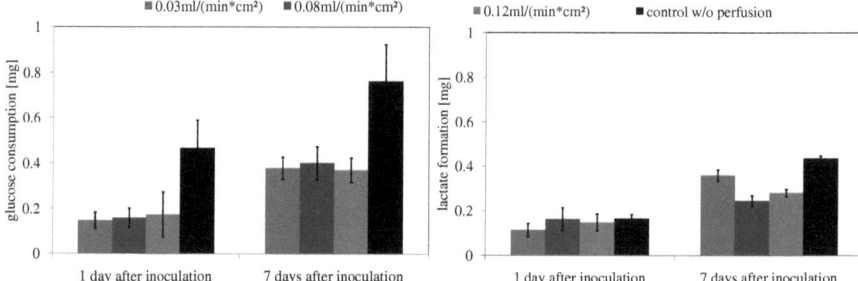

Figure 6.8: Single foam analysis for ceramics with blind holes statically inoculated with *CHO-K1* adapted to growth in medium with reduced serum content 1% FBS ($1 \cdot 10^6$ cells/foam) and followed by cultivation under perfusion conditions for 1 or 7 d (shown is data from a 3 h static cultivation of ceramics in wells of culture plates with fresh medium). Depicted are mean values for glucose consumption (left) and lactate formation (right) of all seven foams of one reactor module (n=7) and of three foams of the control group (n=3) which was cultivated under static conditions in culture plates.

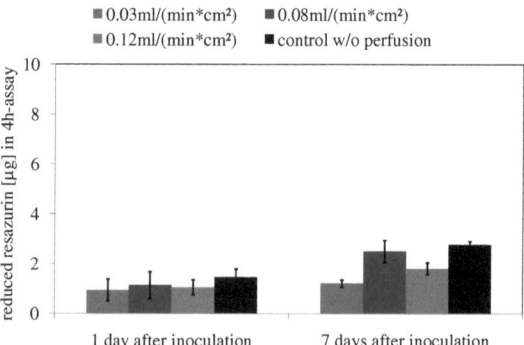

Figure 6.9: Reduction of resazurin (4 h assay) for single foams with blind holes after static inoculation with *CHO-K1* adapted to growth in medium with reduced serum content 1% FBS ($1 \cdot 10^6$ cells/foam) and perfusion cultivation. Depicted are mean values of all seven foams of one reactor module (n=7) and of three foams of the control group (n=3) which was cultivated under static conditions in culture plates.

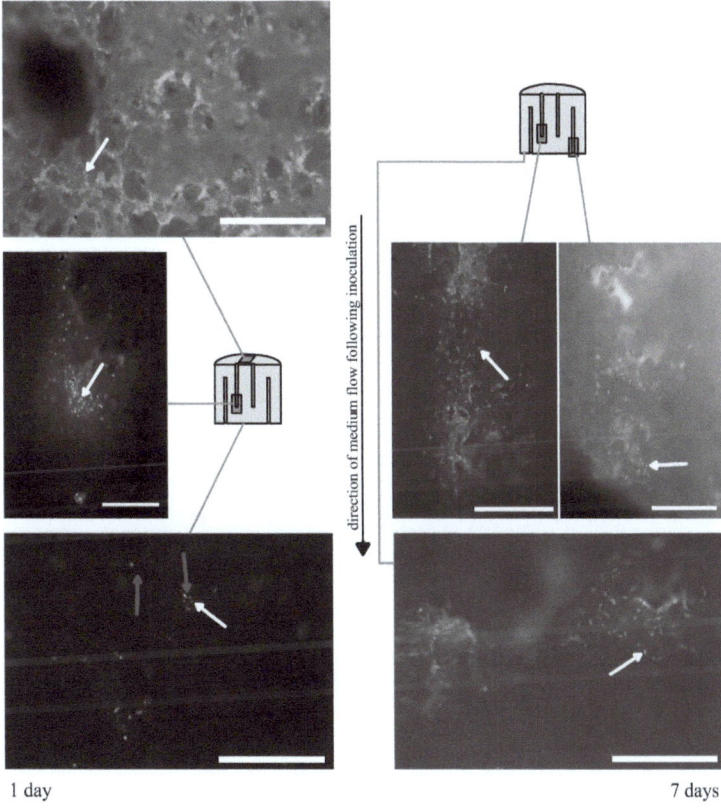

Figure 6.10: Vitality assay for ceramics with blind holes statically inoculated with *CHO-K1* adapted to growth in medium with reduced serum content 1% FBS ($1 \cdot 10^6$ cells/foam) and cultivated for 1 and 7 days at $0.08\,\text{ml min}^{-1}\,\text{cm}^{-2}$. White arrows point to single vital cells, red arrows to dead cells, bars represent 500 µm.

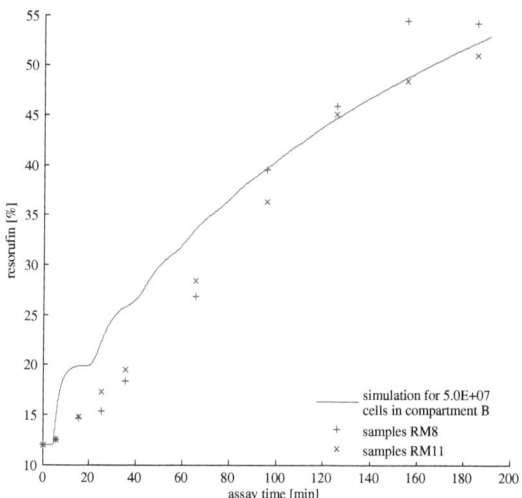

Figure 6.11: Percentage of reduced resazurin in a dynamic resazurin assay for *CHO-K1* statically inoculated into a revolver reactor ($1 \cdot 10^7$ cells/module, 6 foams/module) after 8 days of perfusion cultivation. Revolver module RM10 was eliminated from analysis due to experimental obstacles during resazurin assay. Solid line presents data of a simulation as indicated in the legend.

for the experiments above, data obtained during medium perfusion as in figure 6.7 should be regarded as more trustworthy than data from static assays as in figure 6.8 and 6.9 if the total foam volume is considered. However, in later experiments, I implemented the resazurin assay into the reactor module in order to allow for cells growing inside the ceramic matrix.

6.1.5 Reproducible cultivation following static inoculation into revolver reactors

To prove repeatable operation procedures, another experiment was designed for static inoculation. Three revolver reactors (RM8, RM10 and RM11) with conical magazines (see figure 3.6b) were equipped with six ceramic foams and inoculated each with $1 \cdot 10^7$ *CHO-K1* cells as in chapter 3.3.1. Cultivation was performed for 8 days under perfusion conditions with $0.33 \,\mathrm{ml\,min^{-1}\,cm^{-2}}$ as described in chapter 3.4.1. The reactor modules then were subjected to a resazurin assay as in chapter 3.6.5. Data of dynamic resazurin reduction was close to that of a simulation with $5.0 \cdot 10^7$ cells in reactor modules (chapter 5.1.3), see figure 6.11. Therefore, I assume total cell number has increased about fivefold in ceramics during cultivation of 8 days. Medium samples were further analyzed for glucose consumption and lactate formation as in chapter 3.6.4. Figure 6.12 gives an overview of metabolism, which was very homogeneous for all three reactor modules. Again, microscopic evaluation of single foams by FDA/EB staining revealed a dense layer of cells on the foam's upsides with very few cells at the bottom and no cells inside the ceramic. Moreover, by scanning electron microscopy as in chapter 3.6.3, I did

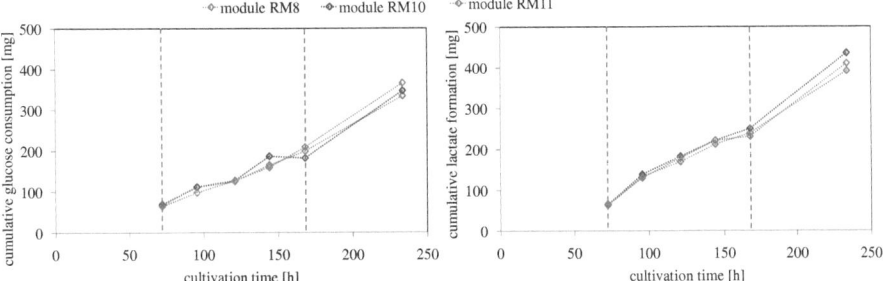

Figure 6.12: Cumulative glucose consumption (left) and lactate formation (right) for *CHO-K1* statically inoculated into cone-magazine reactor modules ($1 \cdot 10^7$ cells/module). Perfusion was performed for 8 days at $0.33 \, \text{ml min}^{-1} \, \text{cm}^{-2}$. Dashed horizontal lines indicate medium replacement.

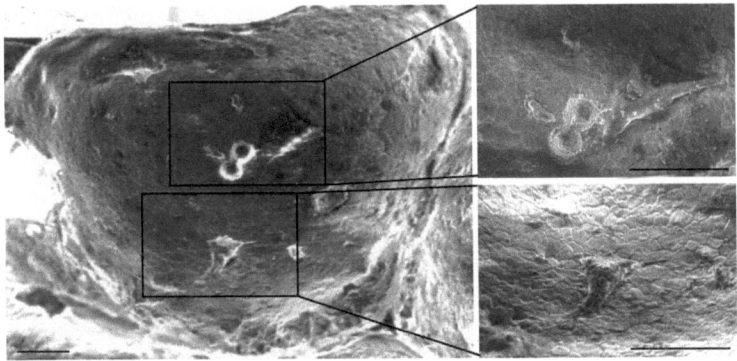

Figure 6.13: Scanning electron microscopy of *CHO-K1* on ceramic foams following static inoculation. Images are acquired in 1 mm depth of the ceramic from above after 8 days of perfusion cultivation. Bars represent 20 µm.

not find many cells inside the ceramic, see figure 6.13.

Static inoculation - synopsis

Static inoculation of *CHO-K1* in all experiments described above led to a dense layer of cells on the foams' surface. Cells proliferate on the ceramic surface (figure 6.1 and 6.9), but growth to higher cell numbers is prevented, probably due to effects such as the community effect [Li et al., 2001]. Cells were found on the foams top and bottom with a mucous matrix covering foams and magazines (figure 6.4), but cellular ingrowth into the three dimensional structure did not occur - which is in good agreement with previous findings outlined in chapter 2.1.3. For static inoculation followed by perfusion cultivation in tubular or revolver reactors, dense layers of cells might even prevent medium from flowing through the scaffold. This is supported by the experiments described in chapter 6.1.2, where cells seemed to pass the inoculated top foam

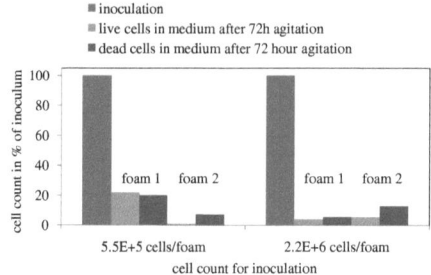

Figure 6.14: Cell absorption by agitation assisted dynamic inoculation of foams in a cell solution. Shown is the cell count in medium solution following 72 h of agitation inoculation for two foams for each inoculation density.

laterally and therefore were able to populate top and bottoms of the following foams. Clearly, if medium does not pass through the foam, cellular ingrowth is inhibited. As shown in chapter 6.1.3 and 6.1.4 there is an optimal flow velocity for the cultivation of *CHO-K1* following static inoculation, which not only determines cellular growth but also affects cellular ingrowth into the matrix. Low flow rates correspond to poor nutrient and oxygen supply and therefore promote cell death. High flow rates promote cellular ingrowth into the ceramic, but also correspond to a decrease in total cell number inside the foam and decreased cell viability, probably due to higher shear stresses. For reactor modules with plain foam holding magazines, I found flow velocities of 0.29 or 0.33 ml min^{-1} cm^{-2} to support growth best following static inoculation and velocities not higher than 0.1 ml min^{-1} cm^{-2} for static inoculation of serum reduced *CHO-K1* into ceramics with blind holes. Nevertheless, proliferation in foams in reactor modules was not as high as in statically cultivated foams, where 8 day cultivation led to an increase of cell number by factor 8 to 14 (chapter 6.1.1) compared to only fivefold by perfusion cultivation (chapter 6.1.5).

Reproducible cultivation of statically inoculated cells in a perfusion reactor is possible, but cell distribution is weak and thick layers of cells on the foam's surface are made up of cell aggregates which will eventually die due to bad nutrient supply as shown by [Sutherland et al., 1986]. Moreover, perfusion cultivation does not enhance cellular growth and therefore is unnecessary, as proven by static cultivation experiments (see figure 6.8).

6.2 Dynamic inoculation by agitation

For agitation inoculation as in chapter 3.3.2 *CHO-K1* with $5.5 \cdot 10^5$ and $2.2 \cdot 10^6$ cells per foam, respectively, were inoculated in 15 ml medium and agitated for 72 h. Cells in medium were collected and centrifuged to evaluate number of live and dead cells using a hemocytometer (see Appendix). Figure 6.14 and table 6.1 summarize the results of dynamic inoculation using agitation. Percentage of cells in medium after 72 h of agitation was much lower for inoculation with $2.2 \cdot 10^6$ cells/foam than for $5.5 \cdot 10^5$ cells/foam, therefore, more cells are assumed to have

Table 6.1: Agitation inoculation of different cell numbers of *CHO-K1* on ceramic foams. Whereas seeding efficiency (calculated as in equation 2.2) strongly differs, optical evaluation did not show remarkable differences in cell number and distribution on the ceramics, see figure 6.15.

cell number for inoculation	total dead cell count in medium after 72 h of agitation in % of inoculum	calculated inoculation efficiency
$5.5 \cdot 10^5$ (foam 1)	20	0.6
$5.5 \cdot 10^5$ (foam 2)	7	0.9
$2.2 \cdot 10^6$ (foam 1)	5	0.9
$2.2 \cdot 10^6$ (foam 2)	13	0.8

adhered to the ceramic surface for the higher inoculum. Nevertheless, foam dyeing as in chapter 3.6.2 did not reveal differences in overall cell density on the foams. Here, cell distribution was inhomogeneous for the foams' surface area with no cells inside the ceramics, see figure 6.15.

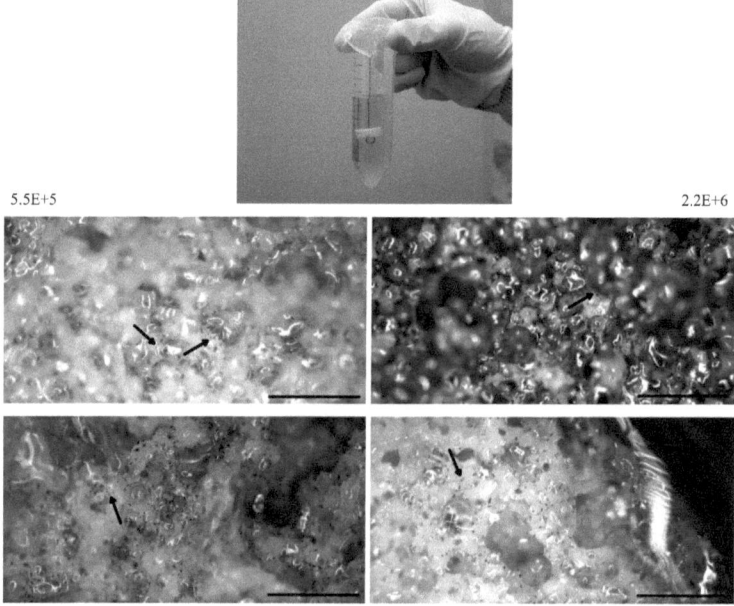

Figure 6.15: Cell density and distribution of *CHO-K1* following agitation seeding. Left: Inoculated cell number $5.5 \cdot 10^5$, right: $2.2 \cdot 10^6$ cells/foam inoculum. Upper pictures show the top, lower pictures the bottom of the ceramic, foams were aligned for agitation inoculation as in picture on top. Bars represent 500 µm, arrows point to single cells.

module alignment	initial cell count/-foam	cycle count	flow velocity [ml min^{-1} cm^{-2}]	cell vitality / cell distribution
module horizontically aligned with rotation*	$3.5 \cdot 10^6$	3.5	0.33	- -/-
			0.24	· / ·
module vertically aligned with horizontal holding*	$2 \cdot 10^6$	3.5	0.24	++/+
	$2.5 \cdot 10^6$			++/++
module vertically aligned	$3.5 \cdot 10^6$	3.5	0.33	-/ ·
			0.24	++/+
			0.19	· / ·
	$5 \cdot 10^6$	7	0.24	++/+
	$2.5 \cdot 10^6$	4.5	0.24	++/++
	$2.5 \cdot 10^6$	3.5	0.33	++/++

Table 6.2: Optical analysis of *CHO-K1* on foams two days after dynamic perfusion inoculation in reactor modules. Cell vitality is expressed as ++ for more than 90% and + for 80-90% vital cells, as · for 70-80%, - for 50-70% and - - for less than 50% vital cells. Cell distribution is expressed as - for no cells in the foam's interior, · for much less cells inside the foam than on the upper and lower face area, + for cells all over the foam's volume with colony forming, and ++ for a homogeneous cell distribution all over the foam's volume (*: see chapter 3.3.3 for further explanation).

6.3 Oscillatory perfusion inoculation

6.3.1 Influence of module orientation, flow velocity, initial cell count on cell distribution

For perfusion assisted inoculation as in 3.3.3, experiments were performed whereby module orientation, initial cell count, number of cycles and flow velocity were varied, see table 6.2. The upper glass olive volume was set to 10 ml prior to inoculation. Data as evaluated in table 6.2 is selected from a wider range of experiments with experiments omitted when results did not lead to additional information. The results, see table 6.2, reveal a strong influence of initial seeding density and initial flow velocity on cell vitality. Cell vitality was very low for the 0.33 ml min^{-1} cm^{-2} /3.5 · 10^6 cells/foam experiments, but very high for the combination 0.33 ml min^{-1} cm^{-2} /2.5 · 10^6 cells/foam when the module is vertically aligned. A flow velocity of 0.24 ml min^{-1} cm^{-2} with 2.5 · 10^6 cells/foam initial seeding density also yielded good results for both vertical alignment and vertical alignment with horizontal holding, whereas higher cell densities often were found to result in low vitality.

Moreover, initial seeding density played an important role regarding even distribution of cells. For the system at hand it should be around 2.5 · 10^6 cells/foam to get cells distributed all over the foam's volume. Both cycle count and module orientation did not influence cell vitality or distribution within the tested frame.

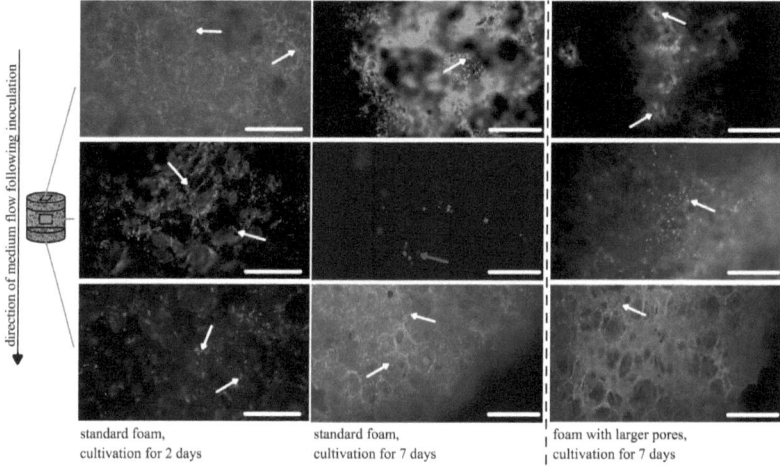

Figure 6.16: Fluorescent staining of *CHO-K1* cells on ceramic foam slices subjected to dynamic seeding as in chapter 3.3.3 with initial seeding density of $2.5 \cdot 10^6$ cells/foam, 3.5 cycles and seeding flow velocity $0.33\,\text{ml\,min}^{-1}\,\text{cm}^{-2}$. Left: Ceramic foam two days after dynamic inoculation. Middle: Ceramic foam seven days after dynamic inoculation. Right: Ceramic foam with larger pores seven days after dynamic inoculation with augmentation of flow velocity (see chapter 6.3.2). Green fluorescence corresponds to vital cells, red fluorescence to dead cells, bars represent 500 µm, arrows point to single cells (white arrow, live cell; red arrow, dead cell); images in the middle row are aquired in approximately 5 mm depth.

I found best results for a seeding density of about $2.5 \cdot 10^6$ cells/foam and an inoculation velocity of 0.24 or $0.33\,\text{ml\,min}^{-1}\,\text{cm}^{-2}$, with 3.5 cycles and vertical alignment of the module as most convenient parameters. Hereby cells were distributed uniformly all over the scaffold's volume, cell vitality was around 100% and cell contact was found to be close without cell aggregates after two days of cultivation. For the higher flow of $0.33\,\text{ml\,min}^{-1}\,\text{cm}^{-2}$ cultivation for another five days did not affect cellular distribution but I observed enhanced cell death in the foams caverns, see figure 6.16 (second column), presumably due to bad nutrient supply. Therefore, initial flow velocity was chosen to be at least $0.33\,\text{ml\,min}^{-1}\,\text{cm}^{-2}$ for the following experiments.

On both top and bottom, scanning electron microscopy as in chapter 3.6.3 revealed cellular growth and additional fibrous structures, presumably matrix proteins, see figure 6.17.

6.3.2 Introduction of more porous ceramics and augmentation of flow velocity during cultivation

It was assumed that poor nutrient supply in 7-day-experiments was attributed to small pore sizes and a flow velocity too low to support the growing cells, therefore in another experiment I introduced a more porous ceramic with mean pore radius approximately 50 µm higher. Ceramics hereby were fabricated as in chapter 3.1 except all substances were milled for 45 min instead

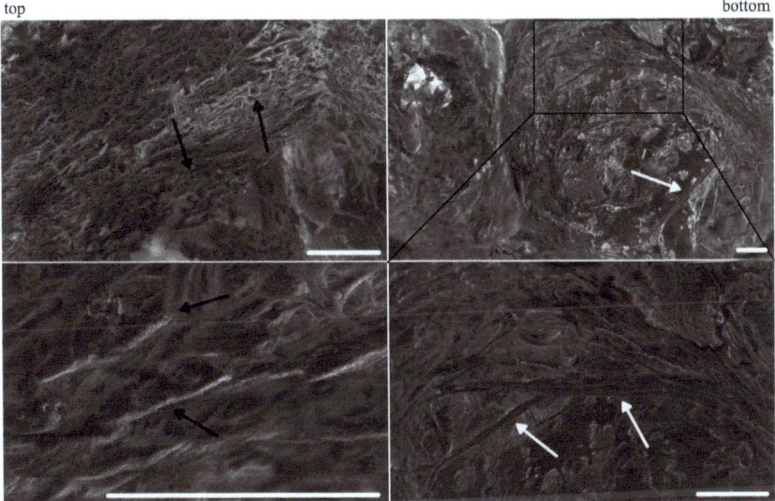

Figure 6.17: Scanning electron microscopy of *CHO-K1* cells on ceramic foam subjected to dynamic seeding as in chapter 3.3.3 with initial seeding density of $2.5 \cdot 10^6$ cells/foam, 3.5 cycles and seeding flow velocity $0.33\,\text{ml}\,\text{min}^{-1}\,\text{cm}^{-2}$. Bars represent 40 µm, black arrows point to single cells , white arrows to fibrous structures observed all over the scaffold. Left: View on top of ceramic foam seven days after dynamic inoculation. Right: View on bottom of the same foam.

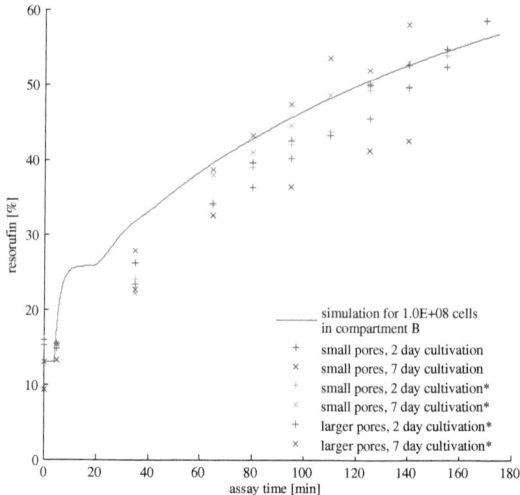

Figure 6.18: Comparison of reduction of resazurin by *CHO-K1* on ceramics with different pore sizes dynamically inoculated and cultivated in reactor modules with augmentation of flow velocity. * labels experiments for which flow was augmented from 0.33 to 0.4 ml min^{-1} cm^{-2} during cultivation on day 5. Solid line presents data of a simulation as indicated in the legend.

of 15 min in the first step. Moreover, flow velocity was augmented during bioreactor operation to ensure sufficient nutrient supply. Three experiments were performed, one with small pore ceramics (standard foams) and constant perfusion flow, one with standard foams and flow augmentation and another with larger pores and also flow augmentation. Cells were inoculated as before in vertically aligned modules with conical magazines with an upper glass olive volume of 10 ml and $2.5 \cdot 10^6$ cells/foam, 3.5 cycles and seeding flow velocity 0.33 ml min^{-1} cm^{-2} followed by an adhering time of 30 min before perfusion started. Flow was augmented to 0.4 ml min^{-1} cm^{-2} on day 5 and ceramics were analyzed following 2 and 7 days, respectively, by a resazurin assay performed inside the reactor module as in chapter 3.6.5 and vitality dyeing as in chapter 3.6.1. I thereby found 100% vital cells following seven days of cultivation only in the foams caverns with higher mean pore radius (figure 6.16, last column), and an increase in metabolic activity compared to constant flow velocities and small pores as evaluated by resazurin assay, see figure 6.18, but not for glucose consumption and lactate formation (chapter 3.6.4), see figure 6.19. From a simulation of resazurin reduction as in chapter 5.1.3, cell number was estimated to be around $1.0 \cdot 10^8$ (corresponding to about $2 \cdot 10^7$ cells per ml foam volume), thereby having increased by factor ≈ 7 for ceramics with larger pores. Main proliferation hereby occurs during the first two days, as resazurin reduction is not much higher 5 days later, see figure 6.18. For small pores, reduction following 7 days of cultivation is even lower than for 2 day cultivations, which is in good agreement with cell depletion observed in vitality assay, see also first two columns in figure 6.16.

6. Influence of mode of inoculation on cellular growth and distribution

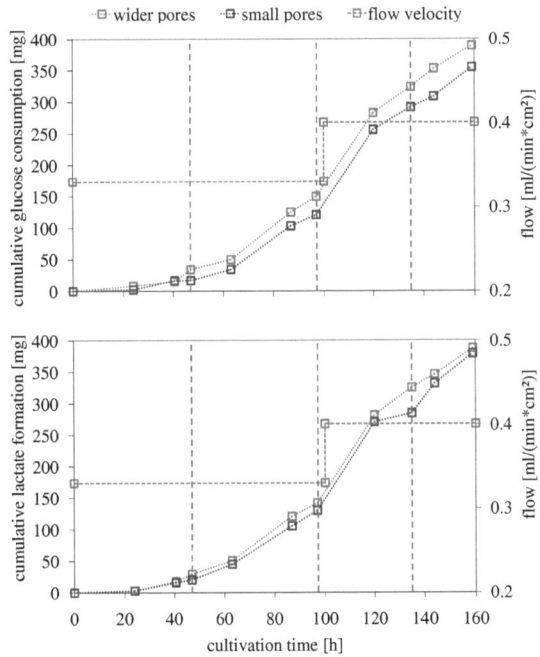

Figure 6.19: Cumulative glucose consumption and lactate formation for *CHO-K1* dynamically inoculated in reactor modules equipped with ceramics with small and larger pores ($2.5 \cdot 10^6$ cells/foam, 6 foams/module) and cultivated with augmentation of flow velocity during cultivation. Dashed horizontal lines indicate medium replacement.

denomination	initial cell count/foam	initial volume in upper glass olive [ml]	top	cross section	bottom
RM8	$1.3 \cdot 10^6$	10	++	o	o
RM9	$3.8 \cdot 10^6$	10	+	+	o
RM11	$2.5 \cdot 10^6$	15	o	o	o
RM12	$1.3 \cdot 10^6$	20	+	+	+
RM10	$3.8 \cdot 10^6$	20	++	+	+

Table 6.3: Optical analysis of ceramic foams with reduced height two days after dynamic perfusion inoculation. Cell vitality in all experiments was found to be 100%. Cell distribution is expressed as o for many inhomogeneously distributed cells, + for a homogeneous cell distribution, and ++ for a dense layer of cells.

6.3.3 Reduction of foam volume by reducing cylinder height

In all perfusion inoculation experiments above, cellular ingrowth in depths of about 2 mm was observed to be very homogeneous from both sides of the foam. Depending on operation strategies and foam geometry, ingrowth could be extended to 5 mm from both sides but cell density was found to decrease with depth. Therefore, I introduced a ceramic, which is only 5 mm in height, hoping for homogeneous growth all over the scaffold.

6 ceramics with larger pores as described in chapter 6.3.2 were assembled into a cone magazine bioreactor and prepared for cultivation. Inoculation was performed with three different cell numbers and three starting volumes in the upper glass olive as in table 6.3 at $0.33\,\mathrm{ml\,min^{-1}\,cm^{-2}}$ and 3.5 cycles. Following 30 min of adhering time, perfusion started from above with $0.33\,\mathrm{ml\,min^{-1}\,cm^{-2}}$ and was performed for two days. Then, foams were subjected to resazurin assays inside the reactor modules and further analyzed in a vitality assay or prepared for scanning electron microscopy as in chapter 3.6.3.

Regarding metabolism, all modules showed a similar behavior except RM8 which was inoculated with the smallest number of cells in the smallest volume, see figure 6.20. For $2.28 \cdot 10^7$ cells/module (RM9 and RM10), there was no difference in metabolism for different inoculation volumes. Vitality assay moreover revealed the most homogeneous cellular distribution for the combination $0.76 \cdot 10^7$ cells/module and 20 ml initial glass olive volume (RM12), see table 6.3 and figure 6.21. Simulation of the resazurin reduction model in chapter 5.1.3 revealed a number of $4.0 \cdot 10^7$ cells per module at maximum (see data for RM12 in figure 6.20), which means an increase in cell number by factor ≈ 5 for just two days of cultivation for RM12 corresponding to exponential growth with a doubling time of 20 h. All in all, flat foams with larger pores yielded very good results regarding cellular distribution and metabolic activity. Here, initial seeding density plays an important role and has to be adjusted carefully.

Oscillatory perfusion inoculation - synopsis

Clearly, regarding cell distribution, dynamic inoculation yields best results. Whereas for inoculation procedure module alignment and cycle count do not influence cell distribution, there is

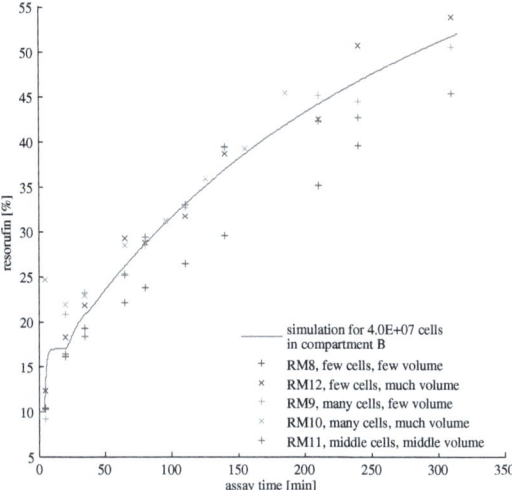

Figure 6.20: Reduction of resazurin for ceramics reduced in height and with larger pores. *CHO-K1* were dynamically inoculated into reactor modules as in table 6.3 and cultivated for two days. Resazurin assays were performed inside the reactor devices. Solid line presents data of a simulation as indicated in the legend.

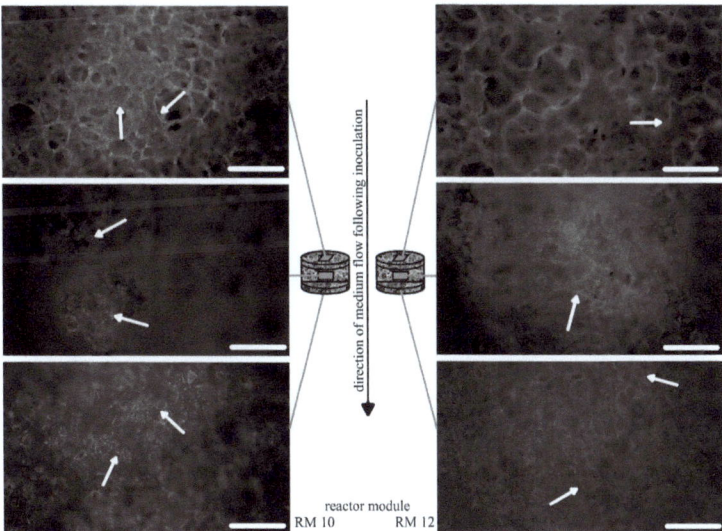

Figure 6.21: Vitality assay for *CHO-K1* on ceramic foams reduced in height and with larger pores two days after dynamic inoculation. Shown are representative pictures of modules inoculated with $3.8 \cdot 10^6$ cells per foam (RM 10) and $1.3 \cdot 10^6$ cells per foam (RM 12) as indicated in table 6.3. Images in the middle row are aquired from cross sections in 2.5 mm depth. Bars represent 500 µm, arrows point to single cells.

a strong influence for initial seeding density and initial flow velocity (see table 6.2). Remarkably, initial cell distribution following dynamic inoculation is not compellingly maintained, but fluctuates with cultivation time. Whereas cell vitality inside the ceramic can be improved by augmenting flow velocities, a much better experimental outcome is achieved when ceramics with larger pores are used (chapter 6.3.2).

In summary, operation procedures for standard foams should be inoculation performed with $1.5 \cdot 10^7$ cells/module holding a conical magazine with 6 foams, upper glass olive volume should be 10 ml, flow velocity 0.33 ml min^{-1} cm^{-2} with 3.5 cycles at vertical alignment. For ceramics with larger pores and reduced in height, inoculation should be performed with $7.6 \cdot 10^6$ cells/module (also holding a conical magazine with 6 foams) in 20 ml upper glass olive volume, 3.5 cycles and 0.33 ml min^{-1} cm^{-2} flow velocity at vertical alignment. During further cultivation, flow velocity should be augmented approximately every 5 days.

7 Reproducibility of the chosen operation methods

As discussed on page 72, a set of operation procedures was found that yielded good cellular distribution all over the foam and maintained cells in a vital status. Until then, to minimize costs, experiments were performed with only one to two modules per variation, therefore reproducibility of the established system still had to be proven. Ceramics with larger pores showed better cellular growth and distribution, but due to the not yet completely elaborated fabrication process, there are still issues of material decomposition for this foam geometry. Therefore, reproducibility tests were performed for standard foams as described in chapter 3.1 as well and for ceramics with larger pores.

7.1 Standard foams

Vertically aligned reactor modules with conical magazines and equipped with glass olives were inoculated as described in chapter 3.3.3 with an upper glass olive volume of 10 ml, $1.5 \cdot 10^7$ cells per module, 3.5 cycles and initial flow velocity $0.33 \, \text{ml} \, \text{min}^{-1} \, \text{cm}^{-2}$. Following 30 min adhering time, perfusion started from above at $0.33 \, \text{ml} \, \text{min}^{-1} \, \text{cm}^{-2}$ and cultivation was performed for 14 days. All in all, 10 modules were started at 5 different days. Flow was augmented to $0.38 \, \text{ml} \, \text{min}^{-1} \, \text{cm}^{-2}$ on day five and further to $0.42 \, \text{ml} \, \text{min}^{-1} \, \text{cm}^{-2}$ on day ten. On day 14 a resazurin assay was performed inside the reactor modules and single foams were analyzed for cell vitality (FDA/EB staining). Moreover, 2 foams per module were analyzed in CHGE for carbon content (see chapter 3.6.6).

Figure 7.1 gives an overview for glucose consumption and lactate formation during the course of cultivation. Metabolic status as examined by resazurin assay is delineated in figure 7.2. Disassembling the modules on day 14, I found a thick mucous layer on top of the magazines and all over the inner reactor material. Vitality staining of that matrix was found to contain many cells, not attached to each other but arranged inside the structure. However, cell vitality inside this matrix was lower than 50%, see figure 7.3. Vitality assay of cells on ceramics revealed high vitality all over the foam's volume and good to few cell numbers inside the ceramic scaffolds, see figure 7.3. Cell number as evaluated by equation 5.17 following CHGE was found to be around $4.1 \cdot 10^7$ cells/foam, resulting in $2.5 \cdot 10^8$ cells per module, see table 7.1. As foams were covered with a mucous layer as described above, cell number as evaluated from carbon content analysis clearly is estimated too high, explaining the divergence to cell number

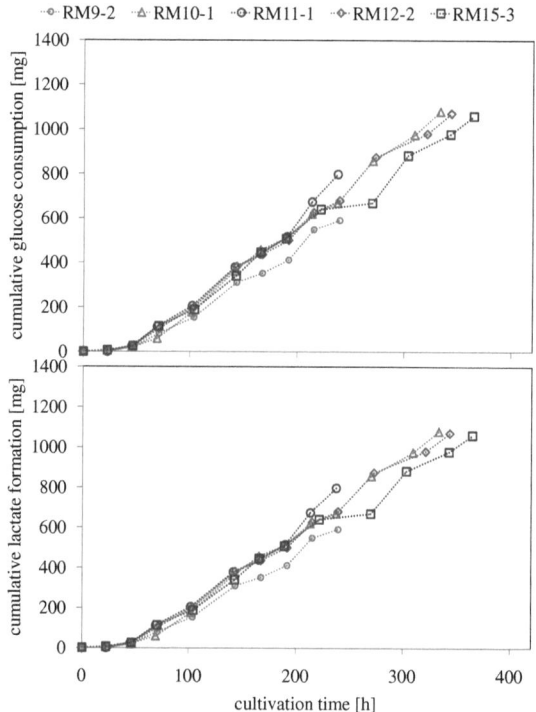

Figure 7.1: Cumulative glucose consumption and lactate formation for *CHO-K1* dynamically inoculated on standard foams with $1.5 \cdot 10^7$ cells per module and cultivated for 14 days with augmentation of flow velocity (see main text, five runs were selected for analysis). Medium replacement was performed every second day.

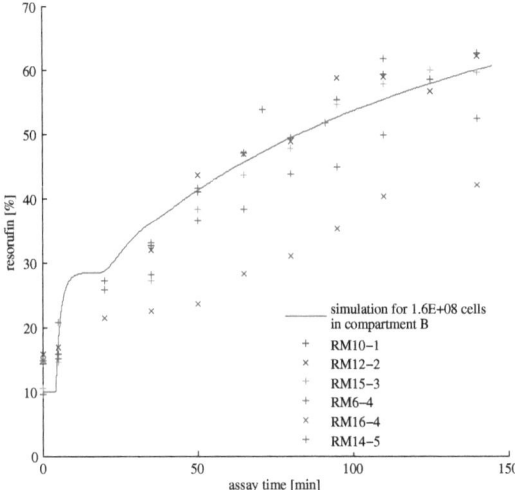

Figure 7.2: Reduction of resazurin for *CHO-K1* dynamically inoculated on standard foams with $1.5 \cdot 10^7$ cells per module and cultivated for 14 days with augmentation of flow velocity (see main text). Shown is data from six modules out of five different runs as indicated in the legend above. Resazurin assays were performed inside the reactor devices. Solid line presents data of a simulation as indicated in the legend.

as estimated from resazurin assays, see table 7.1.

Whereas overall glucose and lactate metabolism gives a very homogeneous picture for all experiments, resazurin assays showed remarkable divergence, see figure 7.2. However, simulation of the resazurin reduction model in chapter 5.1.3 with $1.6 \cdot 10^8$ cells per module was very close to data of some modules (all modules except RM6-4 and RM 16-4 in figure 7.2). Therefore, an increase in cell number by factor of about 11 for 14 day cultivation can be achieved, compared with 7 fold increase following two days of cultivation in figure 6.18.

Nevertheless, out of ten started modules just six could be analyzed properly. Other modules leaked or were contaminated after a medium replacement.

7.2 Foams with larger pores

Vertically aligned reactor modules with conical magazines and equipped with glass olives were inoculated as described in chapter 3.3.3 with an upper glass olive volume of 20 ml, $7.6 \cdot 10^6$ cells per module, 3.5 cycles and initial flow velocity $0.33 \, \text{ml min}^{-1} \, \text{cm}^{-2}$ (conditions as for RM12 in table 6.3). Following 30 min adhering time, perfusion started from above at $0.33 \, \text{ml min}^{-1} \, \text{cm}^{-2}$ and cultivation was performed for 4 days. All in all, 6 modules were started at 2 different days with 6 ceramics with larger pores and 5 mm height each. On day 4 a resazurin assay was performed inside the reactor modules and single foams were analyzed for cell vitality (FDA/EB staining). Moreover, foams were prepared for scanning electron microscopy as in

Table 7.1: Reproducibility of cell cultivation in reactor modules for *CHO-K1* as examined by dynamic resazurin assay and CHGE. (*) Data from resazurin assay is a rough estimation from simulations with different cell numbers as in chapter 5.1.3 and compared to measurement data in figure 7.2 and 7.5.

experiment	cell number per module / SD from resazurin assay*	cell number per module / SD from CHGE
small pores, cylinders 10 mm height, inoculum $1.5 \cdot 10^7$ cells per module, 14 d cultivation (n=6)	$1.4 \cdot 10^8$ / $4 \cdot 10^7$	$2.5 \cdot 10^8$ / $5.1 \cdot 10^7$
wider pores, cylinders 5 mm height, inoculum $7.6 \cdot 10^6$ cells per module, 4 d cultivation (n=6)	$2.0 \cdot 10^8$ / $3 \cdot 10^7$	$9 \cdot 10^7$ / $3.0 \cdot 10^7$

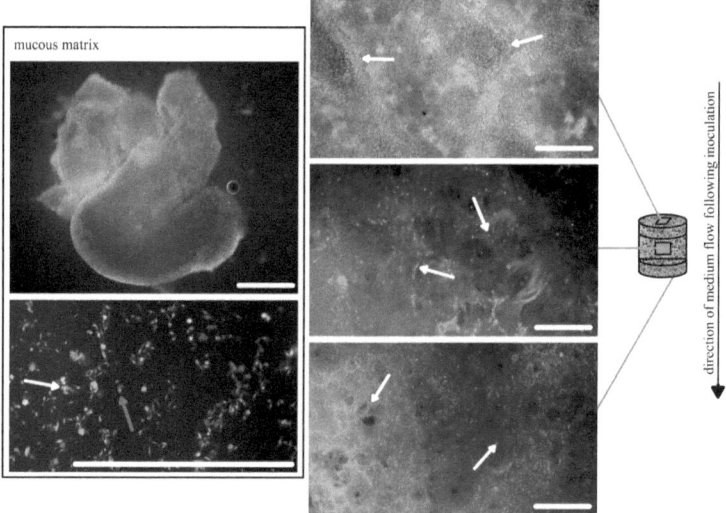

Figure 7.3: Cell distribution for *CHO-K1* dynamically inoculated on standard foams with $2.5 \cdot 10^6$ cells/foam and cultivated for 14 days with augmentation of flow velocity. Left: Cell containing mucous matrix found on top of the foams and inside the reactor modules. Right: Representative images of cell distribution in and on foams after two weeks of perfusion cultivation. The cross section image is aquired in 5 mm depth. Bars represent 500 µm, white arrows point to live cells, red arrows point to dead cells.

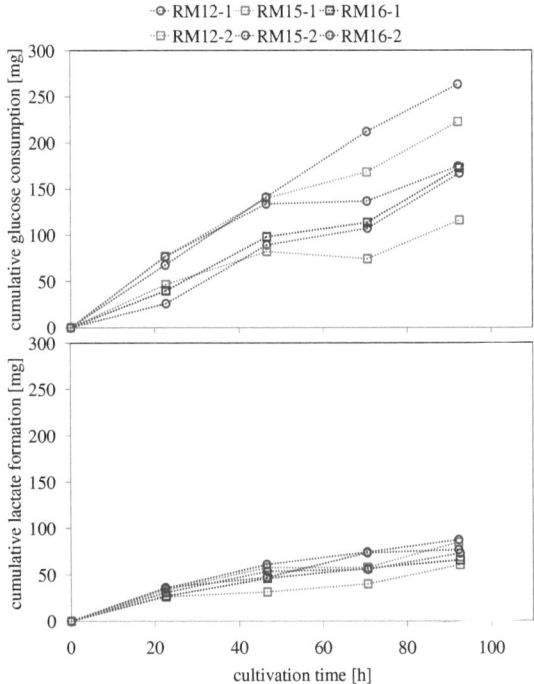

Figure 7.4: Cumulative glucose consumption and lactate formation for *CHO-K1* dynamically inoculated on ceramics with reduced height and larger pores with 7.6 · 10^6 cells per module. Perfusion cultivation was performed for four days.

chapter 3.6.3 and two foams per module were analyzed for carbon content in CHGE (chapter 3.6.6).

Figure 7.4 gives an overview for glucose consumption and lactate formation during the course of cultivation. Metabolic status as examined by resazurin assay is delineated in figure 7.5. Vitality staining revealed high vitality all over the foams' volumes and good to few cell numbers inside the ceramic scaffolds, see figure 7.6. Cell growth was dense on the foams's tops and very homogeneous on the foams' bottoms and inside the foams. Cell density was equal for both bottoms and cross sections but lower than on top of the ceramic. Nevertheless, cells were found homogeneously distributed all over the cross sections with a sufficient cell density for all foams of each reactor module likewise. Scanning electron microscopy revealed dense layers of cells on the foam's surface and fibrous structures in which more cells were observed, see figure 7.7. Cell number as evaluated by equation 5.17 following CHGE was found to be around 1.5 · 10^7 cells/foam, resulting in 9 · 10^7 cells per module, see table 7.1.

For cells inoculated on foams of 5 mm height and with larger pores, cell distribution and metabolic activity as evaluated by resazurin assays was rather homogeneous for all six experiments. However, metabolic data from medium over four days of cultivation showed a broad

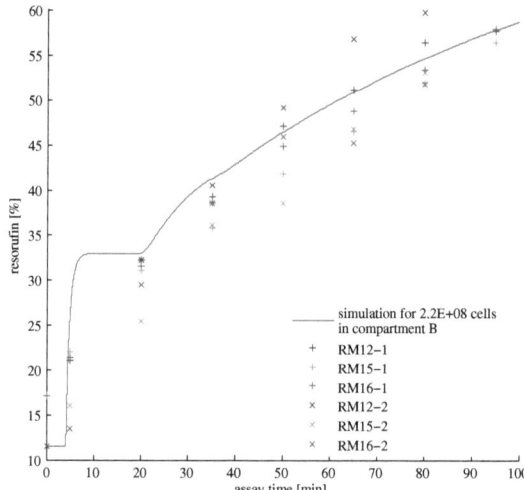

Figure 7.5: Reduction of resazurin for *CHO-K1* dynamically inoculated on ceramics with reduced height and larger pores with $7.6 \cdot 10^6$ cells per module. Perfusion cultivation was performed for four days. Two runs were performed with three reactor modules each as indicated in the legend above. Resazurin assays were performed inside the reactor devices. Solid line presents data of a simulation as indicated in the legend.

distribution (figure 7.4). Compared to cells on standard foams as in chapter 7.1, glucose consumption was around the same depending on the considered reactor module although inoculum was only one half of that for standard foams. Furthermore, reduction of resazurin was even higher for foams with larger pores even though cultivation was performed only for four days compared to two weeks for standard foams and twice as much cells for inoculation. Hereby, cell number was found to have increased by a factor of 29 as evaluated by a simulation as in chapter 5.1.3, see figure 7.5. However, cell number as evaluated by CHGE was found to be much less with a cell incrementation factor of around 12, see table 7.1. Clearly, metabolic assays will take into account cells growing all over the reactor's periphery, whereas CHGE only determines cell number on and inside ceramics. Still, proliferation in ceramics with larger pores is higher, as the incrementation factor for small pores is around 17 (as calculated from CHGE analysis, see table 7.1), but cultivation was performed for 14 days compared to only 4 days for larger pores. Cell distribution all over the scaffold was more homogeneous for foams with wider pores, therefore local cell density was lower, which might have led to an increase in metabolism and proliferation compared to higher foams with smaller pores.

Regarding operation techniques, opposed to modules in chapter 7.1 all six modules were operated sterilely for four days - most likely due to the shorter cultivation time and no needs for medium replacements.

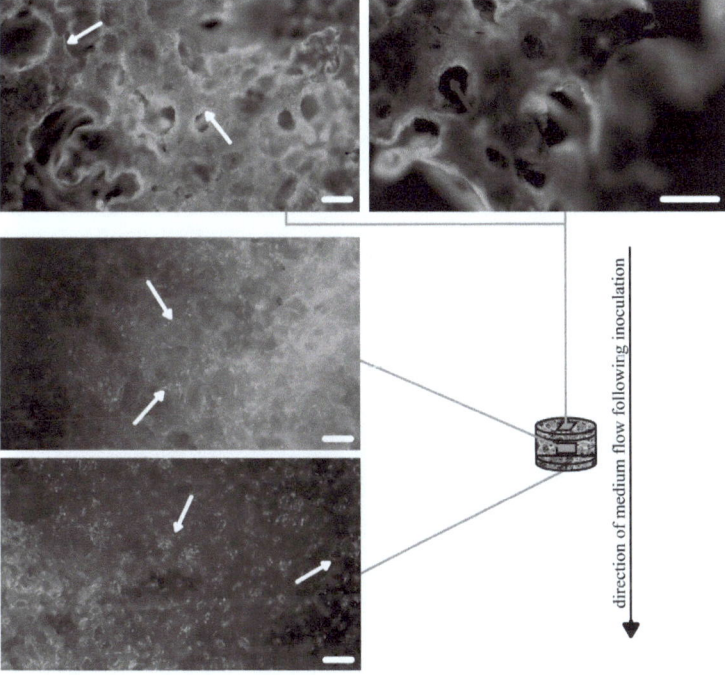

Figure 7.6: Cell distribution for *CHO-K1* dynamically inoculated on ceramics with reduced height and larger pores with $7.6 \cdot 10^6$ cells per module. Perfusion cultivation was performed for four days. Representative images of cell distribution in and on foams cultivated for four days after dynamic inoculation with $7.6 \cdot 10^6$ cells/module are shown. The cross section image is aquired in 2.5 mm depth. Bars represent 200 µm, white arrows point to live cells, blue arrow points to a pore's rim totally overgrown by cells.

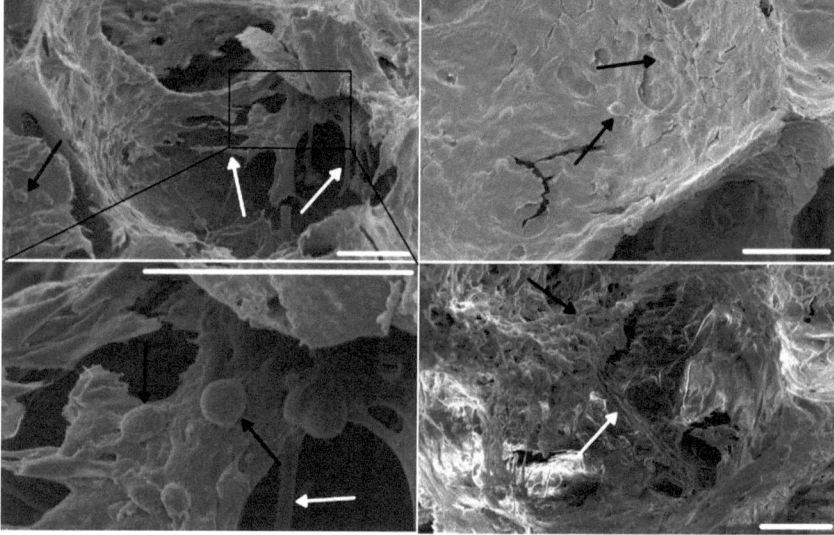

Figure 7.7: Scanning electron microscopy for *CHO-K1* dynamically inoculated on ceramics with reduced height and larger pores with $7.6 \cdot 10^6$ cells per module. Perfusion cultivation was performed for four days. Cells grow on the foam's surface and on fibrous structures, presumably secreted by themselves. Images are taken from the top of the foam and in 2.5 mm depth for image on bottom right. Bars represent 50 µm, black arrows point to single cells, white arrows to fibrous structures.

Reproducibility - synopsis

Cells can be grown more or less reproducible on standard ceramics with operation techniques as found in chapter 6.3.1. Whereas overall metabolism regarding glucose consumption and lactate formation does not vary so much (figure 7.1), cell distribution over the scaffold volume varies strongly between foams of one reactor module, i.e. with a factor 2 for single foam analysis in CHGE following 14 days of cultivation. Moreover, cellular ingrowth into ceramics is limited to 400 µm at maximum and thick mucous layers containing dead cells were found on top of the magazines, see figure 7.3. A better experimental outcome is achieved when cells are inoculated into foams with only 5 mm height and larger pores as in chapter 6.3.3. Here, cell distribution all over the scaffold is very homogeneous and metabolic rates as found in a resazurin assay following 4 days of cultivation do not differ much (figures 7.6 and 7.5). Although total inoculum is smaller, metabolism of cells on foams with larger pores is higher and scanning electron microscopy revealed protein resembling structures with strong cellular ingrowth on the pores' surface (see figure 7.7). From that, as a cell culture scaffold foams with 5 mm height and a coarser pore structure should be favored for cell cultivation in the presented devices. Yet, metabolic data derived from glucose analyzation showed a dispersion of factor 2.5 (see figure 7.4) and single foam analysis in CHGE one of 2, therefore, experimental performance was not quite satisfactory all in all.

8 Long-term cultivation

For many (especially tissue engineering) demands, it is absolutely necessary to maintain cells consistently over a longer cultivation period of a few weeks. Therefore, with operation procedures found in chapter 6.3 four to seven week long experiments were performed to demonstrate the qualification of the established bioreactor system. Cultivations were performed for standard foams as described in chapter 3.1 and for foams with larger pores, respectively.

Two vertically aligned reactor modules with conical magazines and equipped with glass olives were inoculated as described in 3.3.3 with $1.5 \cdot 10^7$ cells and an upper glass olive volume of 10 ml for RM8 and $7.6 \cdot 10^6$ cells and an upper glass olive volume of 20 ml for RM14, 3.5 cycles and initial flow velocity $0.33\,\text{ml}\,\text{min}^{-1}\,\text{cm}^{-2}$. Thereby, reactor module RM8 consisted of 6 ceramix foams of 10 mm height and with standard pore size, the magazine for RM14 contained 6 foams of 5 mm height and with larger pores. Following 30 min adhering time, perfusion started from above at $0.33\,\text{ml}\,\text{min}^{-1}\,\text{cm}^{-2}$ and cultivation was performed for 4 or 7 weeks. Flow was augmented to $0.38\,\text{ml}\,\text{min}^{-1}\,\text{cm}^{-2}$ on day four and further to $0.42\,\text{ml}\,\text{min}^{-1}\,\text{cm}^{-2}$ on day nine (thirteen for RM14). On day 49 a resazurin assay was performed inside the reactor module for RM 8, and single foams were analyzed for cell vitality (FDA/EB staining) on day 28 for RM14 and on day 49 for RM8, respectively. Moreover, two foams each were prepared for scanning electron microscopy as in chapter 3.6.3 and two foams were analyzed for carbon content in CHGE as in 3.6.6 for RM14.

Following 2 weeks of operation, cellular growth was observed in the tubings of the reactors' peripheries with cell clumps of approximately 1-2 mm diameter and cells grew all their way into the medium container during the following cultivation time. Following 4 weeks of cultivation, reactor module RM14 had to be dismantled, as cell growth inside the tubings led to blockage and inhibition of medium perfusion. Therefore, the resazurin assay for RM14 was omitted.

Figure 8.1 gives an overview for glucose consumption and lactate formation during the course of cultivation. Metabolic status for reactor module RM8 as examined by resazurin assay after 7 weeks of cultivation revealed an increase in cell number by a factor of about 24 regarding inoculation density, see figure 8.2. Therefore, overall cell proliferation in ceramics can be assumed to have increased as summarized in figure 8.3.

Vitality staining revealed high vitality all over the foams' volume and good to few cell numbers inside the ceramic scaffolds for RM8. For RM14, cell growth was dense on top of the foams, but no cells were found inside the foams and cell vitality on bottom was only around 90%, see figure 8.4, right. As cell growth was very good for cells cultivated for 4 days (figure 7.6, page 79), the bad experimental outcome here is associated with a poor nutrient supply due to blockage of tubings.

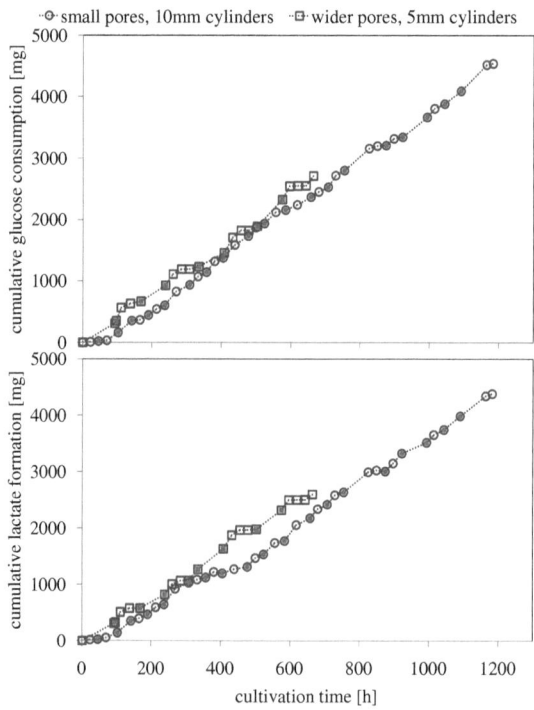

Figure 8.1: Cumulative glucose consumption and lactate formation for *CHO-K1* dynamically inoculated on standard foams with $1.5 \cdot 10^7$ cells per module (circles) and on 5 mm foams with larger pores with $7.6 \cdot 10^6$ cells per module (squares). Cultivation was performed for 49 and 28 days, respectively. Medium replacement is indicated by filled symbols.

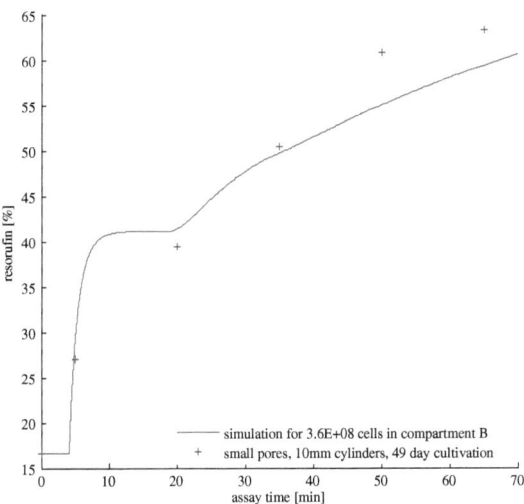

Figure 8.2: Metabolic reduction of resazurin by cells cultivated on standard foams for 7 weeks under medium perfusion ($1.5 \cdot 10^7$ cells/module inoculum). Solid line presents data of a simulation as indicated in the legend.

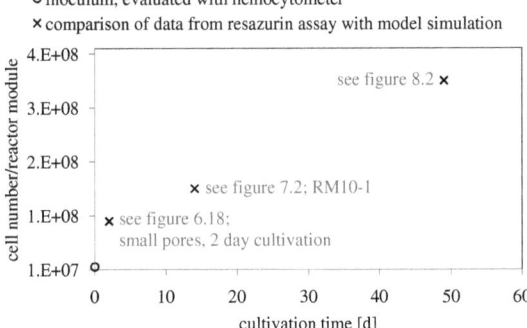

Figure 8.3: Cell proliferation for *CHO-K1* as examined by dynamic resazurin assay for various experiments on standard foams.

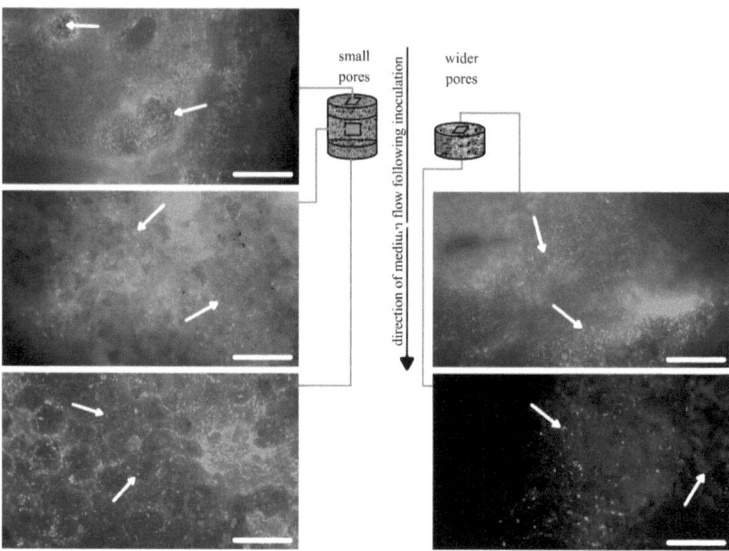

Figure 8.4: Vitality staining for *CHO-K1* cultivated for several weeks in standard foams and foams with larger pores. Left: Analysis of cells inoculated with $1.5 \cdot 10^7$ cells per module on standard foams and cultivated for 7 weeks. Right: Analysis of cells inoculated with $7.6 \cdot 10^6$ cells per module on foams of 5 mm height and with larger pores, cultivation for 4 weeks. Bars represent 500 µm, arrows point to single cells.

Disassembling reactor module RM8 on day 49 revealed a thick mucous layer on top of the magazine and all over the inner reactor material. Again, this matrix was found to contain many cells with about 50% of them dead as in figure 7.3. Cell growth, however, was very good on both top and bottom of the ceramics' with few cells inside the scaffold, see figure 8.4, left. Scanning electron microscopy moreover disclosed abundance of protein like structures all over the foams' volume. For standard foams, cells seemed to grow through that protein matrix, see figure 8.5, left. For RM14, cells were found to grow densely on top of ceramics, see figure 8.5, right, but very sparsely inside the ceramics. As above, due to tubing blocking, nutrient supply is assumed to have been too low to support cellular maintenance inside ceramics for RM14. However, cell number as calculated from equation 5.17 following CHGE was around $1.2 \cdot 10^8$ cells per module, i.e. increased by factor 16 during 4 weeks of cultivation. Obviously, these numbers have to be interpreted very critically, as CHGE determines carbon content from all biological material, which would also be highly abundant mucous matrix. Therefore, total cell number in all ceramics will be substantially lower than estimated by equation 5.17.

Sterile operation of reactor module RM8 with medium replacements twice a week could be performed over seven weeks.

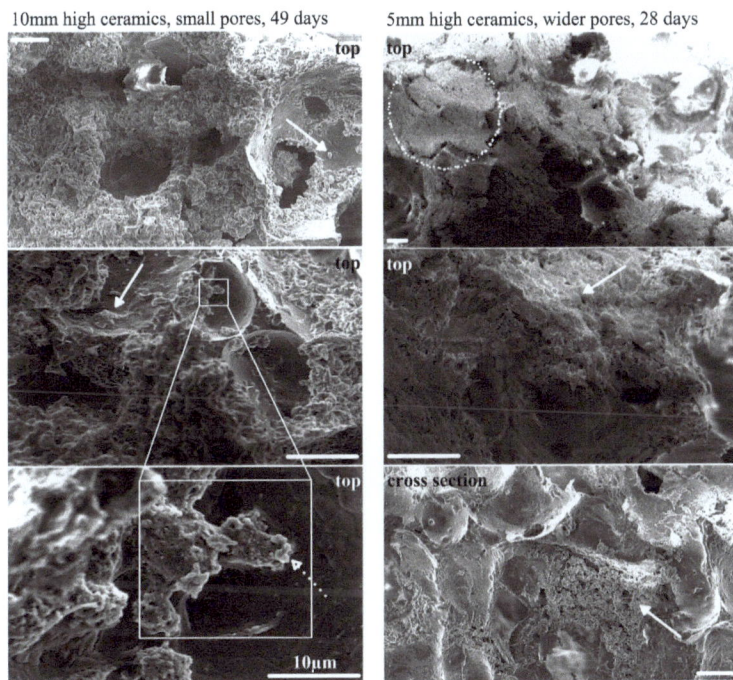

Figure 8.5: Scanning electron microscopy for *CHO-K1* cultivated for several weeks in standard foams and foams with larger pores. Perfusion cultivation was performed for 49 days and 28 days, respectively. Cells grow densely on the foam's surface and on fibrous structures, presumably secreted by themselves. Images are taken from the top of the foam and in 2.5 mm depth as indicated. Bars represent 50 μm unless otherwise stated, white solid arrows point to cells, white dashed arrow to fibrous structures, circle surrounds an area of carpet-like cell growth on the ceramic's surface.

9 Applicability of the reactor system for other cell types

So far, all experiments were performed with *CHO-K1*. Chapter 7 and 8 finally verified reproducible cultivation of *CHO-K1* with good cellular distribution all over the ceramic scaffold and the bioreactor's qualification for long-term cultivation, respectively. Nevertheless, the system was thought to be more or less universal from the beginning. Therefore, three more cell types described in chapter 3.5 were tested for their ability to grow inside the ceramic scaffold in the perfused bioreactor system.

All experiments were performed in reactors with conical magazines as in figure 3.6b with ceramic foam cylinders of 5 mm height and larger pores as in chapter 6.3.3. To find best operational strategies, tests with variations in inoculated cell number, cycle count, initial inoculation volume in glass olives or perfusion velocity were performed.

9.1 Human lung carcinoma cells *A549*

Cells were used in their passage 35 for inoculation. Following injection of cells, the volume of the lower glass olive was pumped through the module into the upper volume at first, then perfusion inoculation continued for another 3.5 cycles. Cells were cultivated with perfusion flow for 3 days before vitality assays (chapter 3.6.1) of populated foams were done. First test runs revealed an influence of initial cell count and flow velocity on cells' distribution and vitality, see table 9.1. Cell growth for the last experiment (IV) was very dense on the foam's top and bottom and less denser for the cross section with still a high number of cells.

For another experiment, cells were cultivated for 14 days. Here, the experimental setup IV from table 9.1 was chosen, as it yielded the best results regarding cell vitality and distribution within the tested range. Cells were optically analyzed by vitality assay and also scanning electron microscopy (chapter 3.6.3) and revealed very good results regarding cell distribution, see figure 9.1. Cell growth on both top and buttom was dense, but single cells still to be recognized. For the cross section, cell distribution was excellent and cell density was good as well. From scanning electron microscopy, single cells were found on the ceramic surface, see figure 9.2.

Table 9.1: Optical analysis of *A549* on ceramic foams with reduced height and larger pores three days after dynamic perfusion inoculation. * stands for the optical analysis of the FDA/EB staining following three days of cultivation for the top and bottom side of the foams and a cross section, respectively. Indicated are cell vitality in % and cell distribution compared to experiment IV, which was set to excellent (exc). o cell distribution is equal to that in experiment IV, - cell distribution is poor compared to experiment IV, and + cell distribution is better.

label	initial cell count/ foam	flow for inoculation $[ml/min \cdot cm^2]$	flow for cultivation $[ml/min \cdot cm^2]$	view from top*	view on cross section*	view from bottom*
I	$1.3 \cdot 10^6$	0.33	0.33	100%/o	100%/-	90%/-
II	$2.5 \cdot 10^6$	0.33	0.33	100%/o	100%/o	100%/-
III	$2.5 \cdot 10^6$	0.66	0.33	100%/+	100%/o	100%/-
IV	$2.5 \cdot 10^6$	0.66	0.66	100%/exc	100%/exc	100%/exc

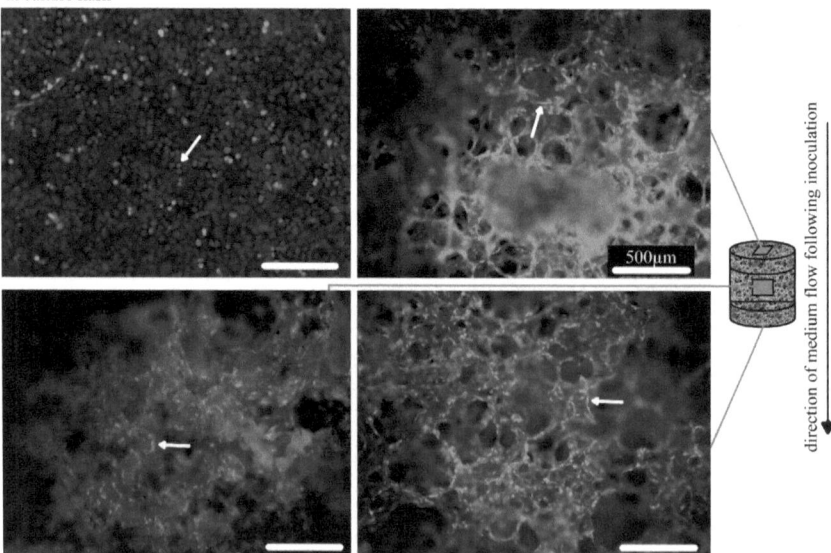

Figure 9.1: Vitality staining of *A549* following 2 weeks of cultivation with setup IV from table 9.1. Top left shows a picture of *A549* grown in a culture flask. The image from the middle of the foam is taken in approximately 3 mm depth. Bars represent 500 µm, arrows point to single cells.

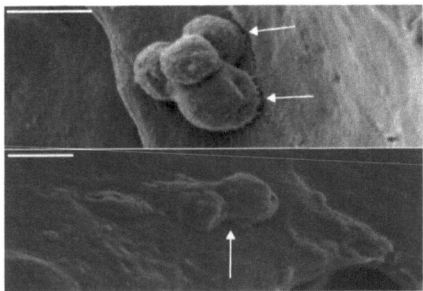

Figure 9.2: Scanning electron microscopy of *A549* following 2 weeks of cultivation, images were taken from the top of the ceramic. Bar represents 10 μm, arrows point to single cells.

Table 9.2: Optical analysis of primary fibroblasts on ceramic foams with reduced height and larger pores 7 days after dynamic perfusion inoculation. * stands for the optical analysis of the FDA/EB staining following cultivation for the top and bottom side of the foams and a cross section, respectively. Indicated are cell vitality in % and cell distribution, which is + for good distribution with dense growth, o for acceptable distribution with single cells and cell clumps occuring, and - for inhomogeneous cell distribution.

label	initial cell count/foam	cultivation duration [d]	view from top*	view on cross section*	view from bottom*
I	$3.3 \cdot 10^5$	7	100%/o	100%/-	100%/o
II	$1.3 \cdot 10^6$	7	100%/+	100%/o	100%/o

9.2 Human primary fibroblasts

Cells were used in their passage 10 for inoculation. Following injection of cells, the volume of the lower glass olive was pumped through the module into the upper volume at first, then perfusion inoculation continued for another 3.5 cycles at $0.33 \, \text{ml} \, \text{min}^{-1} \, \text{cm}^{-2}$. Cells were cultivated at $0.33 \, \text{ml} \, \text{min}^{-1} \, \text{cm}^{-2}$ for 7 days before vitality assay (chapter 3.6.1) of populated foams. As cultivation of cells before inoculation takes very long (see Appendix), only two runs were performed, see table 9.2. For cultivation II of table 9.2, cell growth on both top and buttom was very homogeneous and well distributed and cellular ingrowth was observed for the total foam volume, see figure 9.3. Scanning electron microscopy (chapter 3.6.3) moreover revealed a variety of different structures on the ceramic's surface and between pore connections, which probably are matrix proteins secreted by the fibroblasts, see figure 9.4.

9.3 Madin-Darby canine kidney cells (*MDCK*)

Cells were used in their passage 5 after thawing (see appendix) for inoculation. Following injection of cells, the volume of the lower glass olive was pumped through the module into the upper volume at first, then perfusion inoculation continued for a cycle count as indicated in

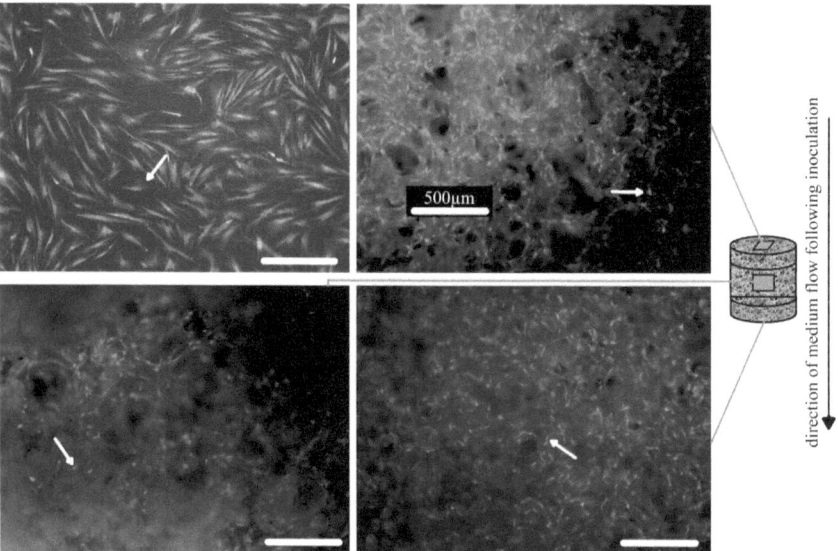

Figure 9.3: Vitality staining of primary fibroblasts following 7 days of cultivation with setup II from table 9.2. Top left shows a picture of primary fibroblasts grown in a culture flask. The image from the middle of the foam is taken in approximately 3 mm depth. Bars represent 500 µm, arrows point to single cells.

9. Applicability of the reactor system for other cell types

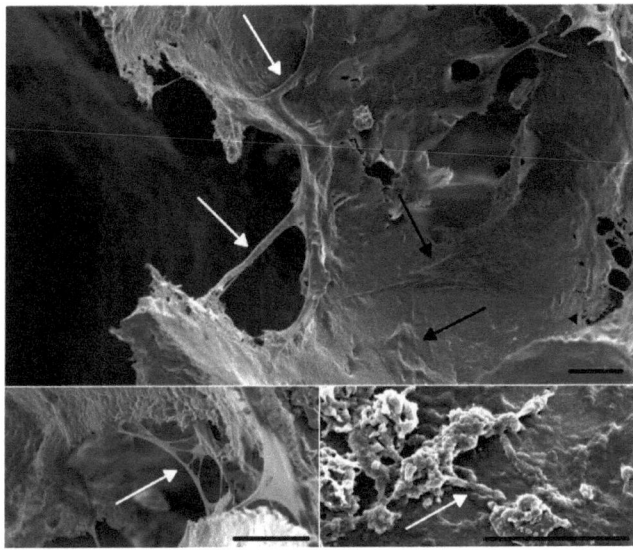

Figure 9.4: Scanning electron microscopy of primary fibroblasts following 7 days of cultivation with setup II from table 9.2, images were taken from the top of the ceramic. Bars represent 20 µm, black arrows point to single cells, white arrows to fibrous structures.

table 9.3 at $0.33\,\mathrm{ml\,min^{-1}\,cm^{-2}}$. Cells were cultivated at $0.33\,\mathrm{ml\,min^{-1}\,cm^{-2}}$ for 3 days before populated foams were dyed as in chapter 3.6.2 and analyzed for cell distribution. The experimental outcome is depicted in table 9.3. For another experiment, cells were cultivated for 14 days. Here, experimental setup IV from table 9.3 was chosen, as it yielded the best results regarding cell vitality and distribution within the tested range. Cells were optically analyzed by hematoxylin/eosin dyeing and also by scanning electron microscopy (chapter 3.6.3), revealing very good results regarding cell distribution, see figure 9.5. Cell growth on both top and buttom was dense, but single cells still to be recognized. For the cross section, cell distribution was excellent and cell density was good as well. However, both cell density and cell distribution decreased with depth, resulting in much less cellular growth on the foams' bottom. Scanning electron microscopy showed a dense layer of cells covering the ceramic surface in the cross section, see figure 9.6.

Introduction of other cell types - synopsis

Growth and high viability of *A549*, primary fibroblasts and *MDCK* was confirmed on the ceramic's top, bottom and inside the ceramic structure for cylinders of 5 mm height and with larger pores compared to standard foams. Here, operation techniques only had to be adjusted regarding inoculation density and flow velocity, whereby only a few experiments had to be performed. Growth and therefore sufficient nutrient supply inside the ceramic was maintained for 14 days. For fibroblasts, the ceramic was partly covered by a layer of fibrous structures (see

Table 9.3: Optical analysis of *MDCK* on foams with reduced height and larger pores three days after dynamic perfusion inoculation. * stands for the optical analysis of hematoxylin/eosin stained ceramics following three days of cultivation for the top and bottom side of the foams and a cross section, respectively. Indicated are cell vitality in % (as evaluated from FDA/EB staining whenever performed) and cell distribution, which is + for good distribution with dense growth, o for acceptable distribution with single cells and cell clumps occuring and - for inhomogeneous cell distribution.

label	initial cell count/foam	cycle count	view from top*	view on cross section*	view from bottom*
I	$8.3 \cdot 10^5$	0.5	+	o	-
II	$8.3 \cdot 10^5$	4.5	+	o	-
III	$2.5 \cdot 10^6$	0.5	+/100%	o/100%	-
IV	$2.5 \cdot 10^6$	4.5	+/100%	+/100%	-
V	$3.6 \cdot 10^6$	2.5	+	o	-
VI	$5 \cdot 10^6$	0.5	o/100%	-/100%	-
VII	$5 \cdot 10^6$	4.5	o/100%	-/100%	-

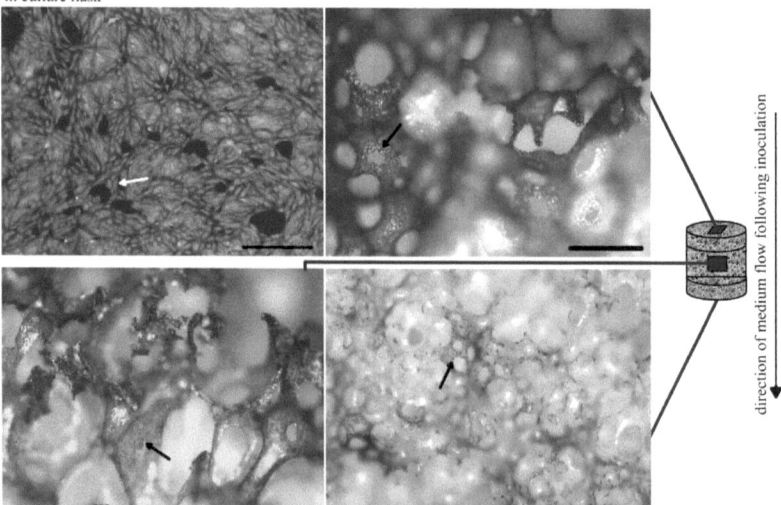

Figure 9.5: Hematoxylin/eosin staining of *MDCK* following 2 weeks of cultivation with setup IV from table 9.3. Top left shows a picture of *MDCK* grown in a culture flask (FDA/EB staining). The image from the middle of the foam is taken in approximately 3 mm depth. Bars represent 500 µm, arrows point to single cells.

9. Applicability of the reactor system for other cell types

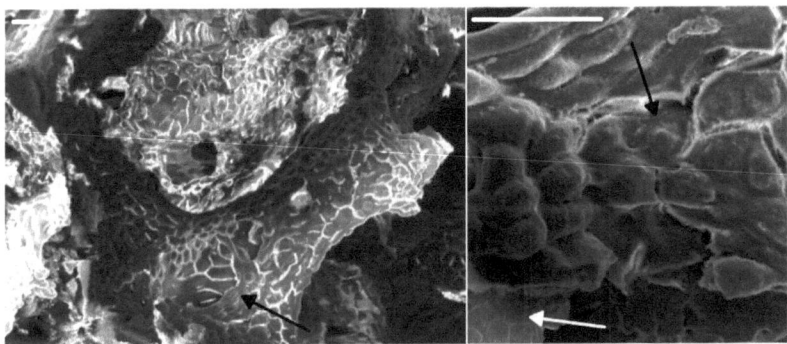

Figure 9.6: Scanning electron microscopy of *MDCK* following 2 weeks of cultivation with setup IV from table 9.3, images were taken from a cross section of the ceramic. Bar represents 20 µm, black arrows point to single cells, white arrows point to the plaster stone-like ceramic surface.

figure 9.4), presumably structural proteins of the ECM as collagen or fibronectin, confirming the hypothesis that cells find a close-to-natural environment inside the porous ceramic structure. For *MDCK* a dense layer of cells was observed even inside the ceramic structure (see figure 9.6) with cells growing on the ceramic's surface and the bridges between pores.

Summarizing, next to *CHO-K1* three further cell types were proven to grow on and in the ceramic structure in the developed bioreactor system.

10 Conclusion

I demonstrated the feasibility of cell cultivation inside porous ceramics by a combined approach of oscillatory inoculation and medium perfusion for cultivation. For this purpose I designed and operated a modular bioreactor system that allows incorporation of ceramics with different geometries, adjustment of multiple operational strategies and allows addition of further components (e. g. glass olives). Four different cell types were cultivated inside the reactor device growing all over the ceramics' surface and volume.

Operation techniques

I found a strong influence of initial inoculation strategy on cell vitality and distribution over the scaffold's volume. Hereby, best results were achieved for seeding cells dynamically in an oscillatory perfusion approach directly inside the reactor device (chapter 6.3) and cultivating them further by means of perfusion, which is in good agreement with findings from the literature [Burg et al., 2000].
To evaluate cell number and metabolism, I established an in-line metabolic assay based on the kinetics of reduction of resazurin (chapter 5.1.3) which allowed the comparison of different experimental outcomes and a first evaluation of cell number inside reactor modules. For an approximation of biomass on and in ceramics I moreover conceived a protocol for the determination of carbon content in ceramics based on carrier hot gas extraction. Carbon content then was used to estimate cell number inside ceramics and hence also for comparison of experiments. A perfusion flow velocity of $0.33\,\text{ml}\,\text{min}^{-1}\,\text{cm}^{-2}$ was found to support growth of *CHO-K1*, *MDCK* and human primary fibroblasts, whereas for *A549* a flow velocity of $0.66\,\text{ml}\,\text{min}^{-1}\,\text{cm}^{-2}$ should be favored. Clearly, due to ceramics' porous structure and therefore tortuosity effects [Habisreuther et al., 2008], actual flow velocity will strongly depend on location inside the scaffold. However, for several flow paths convection will not much be hindered and flow through the porous matrix can be assumed to be $0.33\text{-}0.66\,\text{ml}\,\text{min}^{-1}\,\text{cm}^{-2}$ (corresponding to $0.055\text{-}0.11\,\text{mm}\,\text{s}^{-1}$), which then is in the range of sinusoidal blood flow ($0.01\text{-}0.3\,\text{mm}\,\text{s}^{-1}$) in bone marrow [Wickramasinghe, 1975]. As porous ceramics display strong resemblance with trabecular bone structure hosting bone marrow, diffusion from flow paths with velocities as above is anticipated to be sufficient for nourishment of cells deep inside sparsely accessible pores.

Ceramic scaffold

Regarding foam geometry, a strong influence of length of the ceramics was found, as cylinders of 10 mm height supported cell growth inside the scaffold directly after seeding. However, cell density strongly decreased with scaffold depth and cultivation time (figure 6.16, page 66). Therefore, dynamic oscillatory inoculation was not performed for foams longer than 10 mm or for foams series connected in a tubular reactor as described in chapter 3.2.1. I assume filtration effects very common for porous growth substrates which lead to pore blockage on top of the ceramics and therefore prevent further cell penetration as well as medium from flowing through the total foam volume during further course of cultivation.

For ceramics of 5 mm height I found very good results regarding cell vitality and distribution. Here, total foam volume is occupied by cells (see figure 7.6, page 79). For these more flat foams, pore size was adjusted additionally to be slightly higher than for ceramics of 10 mm height, leading to a coarser foam structure. Good cellular growth in these scaffolds is ascribed to smoother flow through ceramics depending on its shape and internal structure. Therefore, influence of pore size, pore openings and pore edge structure on cellular growth behavior should be studied in more detail to gain deeper insight into cell-ceramic interaction. With hereby gained results, internal pore structure could be adjusted in micro-metre range and prediction and control of cell growth and distribution inside ceramics would be possible. Clearly, feasibility of modulating µm-scaled pore sizes, openings and even flow channels in a scaffold strongly depends on the used material. Therefore, the list of material characteristics suitable for cell cultivation as suggested by [Meuwly et al., 2007] (table 2.1, page 5) should be expanded by the item of good material ductility.

Cell growth

For long-term cultivation, ECM resembling layers were found on top of the ceramics (see figure 7.3, page 76) and fibrous structures were observed inside the ceramics' pores (figure 6.17, page 67). I therefore hypothesize that cells find an excellent micro environment inside ceramics and secrete extra cellular matrix proteins. Nevertheless, following several weeks of cultivation, cell clumps were observed mainly on the ceramics' surface and in the bioreactor's periphery which led to blocking of tubings and a decrease in overall cell vitality as cells inside the aggregates undergo necrosis or apoptosis (chapter 8).

Therefore, especially for week-long cultivations usage of tubings should be avoided and the reactor's interior should be easily accessible to remove cell debris. Timmins and colleagues in a recent publication present a tube-less perfusion bioreactor, the T-CUP [Timmins et al., 2007]. Here, ceramic scaffolds are moved through the medium solution therefore enforcing medium to pass through. As scaffolds are actuated back and forth, not only seeding is performed by help of oscillatory perfusion, but also nutrient supply of cells inside ceramics is ensured by oscillatory fluid flow. Compared to unidirectional flow during cultivation, oscillatory perfusion is known to strongly increase cell vitality inside the scaffold [Du et al., 2008]. In the performed experiments, I found cells which were brought into scaffolds' pores by oscillatory perfusion to die when further cultivated under unidirectional flow (see figure 6.16, page 66). I therefore hypothesize an occurrence of different flow paths through the scaffold for each flow direction. Then, cells in flow paths that can be accessed only from the bottom clearly have to be supplied by medium flow from the bottom of the foam.

Critical remarks

The present essay is a synopsis of quite different fields of science. Data had to be collected for the reactor device (materials and physics, ceramic characteristics, flow patterns), employed assays (cell number determination, mathematical modeling of resazurin reduction, etc.), and bioreactor operation (inoculation, nutrient supply). Herein, besides gained knowledge as how to seal ceramics laterally in order to direct flow through them, several setbacks had to be accepted, e. g., problems in direct assessment of cell number inside the ceramic foams. I however established two methods for cell determination instead, but these are clearly more qualified for comparison of test runs than for specifying explicit cell counts. The resazurin assay, e. g., assumes laminar plug flow behavior inside the reactor device, which clearly is invalid as implied by figure 4.2 (page 35).

I found perfusion inoculation to yield good results regarding cell distribution for ceramics with larger pores, but reproducibility of the experiments was poor (i. e. a dispersion factor of 2.5 for glucose consumption, see figure 7.4 on page 77). Moreover, ceramics with larger pores were found to be less robust, therefore, the manufacturing process should be revised.

Medium flow and hereby nutrient distribution inside the very inhomogeneous ceramic scaffold is a rather random event and probably leads to uneven cell growth. I found mucous layers containing dead cells on top of the scaffolds and inside the reactor periphery in conjunction with unpopulated sites inside the ceramic (see figure 7.3 on page 76), therefore nutrient supply seems to be very inhomogeneous and should be adjusted considering the ceramic's structure.

However, homogeneous cell distributions and growth inside the ceramic scaffolds confirm qualification of the reactor system at hand. Biomasses corresponding to about $4 \cdot 10^7$ cells/ml scaffold volume (as calculated from table 7.1, page 76) were achieved raising confident hope for even more if scaffold structure and nutrient supply are adjusted more elaborately.

Outlook

A huge challenge still needs to be overcome is the abundance of cell aggregates inside the reactor device. Naturally, avoiding tubings - as for the T-CUP mentioned above - only minimizes the risk of bad nutrient supply due to cell blockage but does not hinder cell clump occurrence. Clearly, it would be more useful to prevent the occurrence of cell aggregates in the first place. As cell clumps mainly occur when cell proliferation is high due to nutrient abundance, a well considered strategy of nutrient provision should be incorporated into reactor operation. Hereby, cells would be supplied with just enough nutrients to guarantee cell survival, but further proliferation would be prohibited. Moreover, if deeper insight was gained into how internal structure affects cellular growth (as suggested above), a combined approach of imprinting flow channels and control of nutrient supply could modulate cellular growth spatially.

Recently, other bioreactor systems employing ceramics as cell cultivation scaffolds came up. Here, good results regarding cell distribution over the scaffold, cell viability and even cellular differentiation were shown for cells cultivated under perfusion conditions in scaffolds of a similar shape as those presented in this work [Holtorf et al., 2005, Timmins et al., 2007, Du et al., 2008]. However, trial-and-error-based concepts to improve bioreactor performance are time consuming and expensive. A more elaborate strategy would be to outsource experiments to computations *in silico*, which evidently requires an excellent underlying mathematical model of cellular growth inside porous perfused scaffolds. [Mehta and Linderman, 2005] proved the successful design and operation of a bioreactor depending almost exclusively on theoretical considerations and avoiding laboratory experiments.

All in all, mathematical models describing cellular growth spatially resolved in a porous scaffold would help us to understand cellular growth inside ceramics in more detail, could enable adjustment of favorable micro milieus by nutrient supply and even promote reactor and scaffold improvement.

Then, reactors as presented in this work would represent an excellent basis for three dimensional cell cultivation, comforting cells to behave as in organs and consequently allow fulfilling tissue engineering demands or requirements for production of highly specific therapeutics.

Appendix

List of laboratory equipment

designation	description	origin
shaker	for agitation of ceramics in reaction tubes, works inside a humidified CO_2 incubator	home made design, see figure 10.1
clean bench	for aseptic cell cultivation procedures	LaminAir HB2448, Heraeus Instruments
centrifuge	cell harvesting	Centrifuge 5804R, Eppendorf
autoclave	sterilization of cell culture material/reactor modules	Certoclav, certoclav sterilizer GmbH (Austria)
laboratory oven	drying of ceramics prior to inoculation	T6030, Heraeus Instruments
CO_2 incubator	cell cultivation under defined conditions, 5% CO_2, 90% water saturated air	Heracell 150, Heraeus Instruments
Biostat B plus	serving as a recirculation container for reactor modules	Biostat B plus, Sartorius
climatic chamber	conditioning of modules connected to Biostat B plus	home made design, see figure 3.4
light microscope	evaluation of cell growth for experiments in culture flasks or plates	Diavert, Leitz Wetzlar
upright fluorescence microscope	live/dead differentiation in vitality assay for ceramics	Axio Scope.A1, Zeiss
hemocytometer	determination of cell concentration by microscopic evaluation	Thoma, LaborOptik
photometer	color absorption for resazurin assays and reactor characterization	Tecan Sunrise, Tecan Trading AG

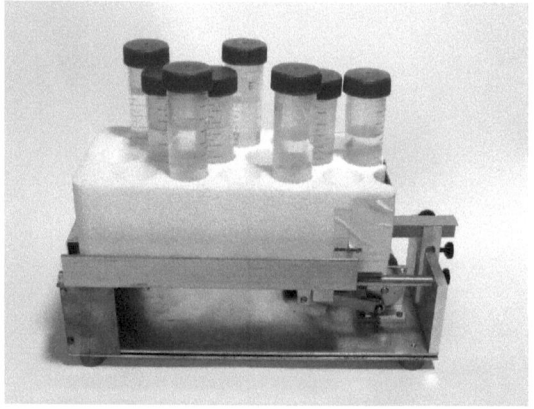

Figure 10.1: Illustration of self-made shaker.

Cell culture materials

designation	description	origin
culture flask	flask for 2-dimensional cell cultivation (growth surface 25, 75 or 175 cm^2	Greiner bio-one, 690175/ 658175/ 660175
cell culture plate	6-, 12- or 24-well plate for 2-dimensional cell cultivation or static cultivation of cells on ceramics	Greiner bio-one, 657160/ 665180/ 662160
reaction tube	50 ml PP tube for agitation cultivation of cells on ceramics	Greiner bio-one, 227261
aqua dest.	double-distilled water	double distilled in Köttermann 1080
PBS	washing solution for cells prior to cell detachment	137mM NaCl (Merck, 1.06404); 2.7mM KCl (Merck, 1.04935); 5.7mM Na$_2$HPO$_4$ (Merck, 1.06580); 1.5mM KH$_2$PO$_4$ (Merck, 1.04873); pH=7.2, in aqua dest.
PBS++	washing solution for cells prior to experiments	137mM NaCl (Merck, 1.06404); 2.7mM KCl (Merck, 1.04935); 4.3mM Na$_2$HPO$_4$ (Merck, 1.06580); 1.8mM KH$_2$PO$_4$ (Merck, 1.04873); 1mM CaCl$_2$ (Merck, 1.02381); 1.5mM MgCl$_2$ (Fluka, 63064); pH=7.2, in aqua dest.
Karnovsky buffer	fixing solution for SEM	0.16M sodium cacodylate trihydrate (Sigma, C0250); 3.6mM CaCl$_2$ (Merck, 1.02381); 0.04% glutaraldehyde (Serva, 23114), in aqua dest.
cacodylate buffer	washing solution for SEM	0.2M sodium cacodylate trihydrate (Sigma, C0250); pH=7.3, in aqua dest.
acetone	dehydration solution for SEM	Riedel-de Haen, 65469
TrypLE Express	cell detaching solution, can be used 3x diluted in PBS	Gibco, 12605

Additional equipment for bioreactor system

designation	description	origin
peristaltic pump	imprinting perfusion to reactor modules by tubings	1B.1003-R/65, Petro Gas Ausrüstungen Berlin
tubings	conducting medium to and from reactor modules, attachment of gas filters to bottles	silicone
gas filter	gas exchange for medium recirculation vessels during cultivation	Acro®50, Pall Corporation
teflon tape	sealing of ceramics	EN 751-3 GRp R10D10, MB Dichtungen GmbH Jetzendorf
glass olives	for dynamic oscillatory cell inoculation	self-made
seal rings	sealing ceramic-PEEK joints, also used as spacers	O-rings coated with 70 EPDM 281
steel spring	separation of ceramics in tube reactor	steel 1.4301 and 1.4571
silicone sealing disks	sealing PEEK-PEEK joints; used as membranes	silicone, shore hardness 60/80
syringe	inoculation of cells through silicone membrane	5 ml, Omnifix B Braun
needle	inoculation of cells through silicone membrane	0.8x80 ml, Erosa

day	passage by split	passage into medium with serum content	cultivation duration
3	1:6	10%	3 d
6	1:6	8%	3 d
9	1:6	8%	2 d
11	1:6	6%	2 d
13	1:6	6%	2 d
15	1:6	4%	2 d
17	1:6	4%	2 d
19	1:6	4%	2 d
21	1:2	3%	3 d
24	1:4	3%	2 d
26	1:6	2%	2 d
28	1:4	2%	3 d
31	1:6	2%	2 d
33	1:6	2%	2 d
35	1:6	2%	3 d
38	1:6	2%	2 d
40	1:6	1%	2 d
42	1:6	1%	3 d
45	1:6	1%	2 d

Table 10.1: Adaptation of *CHO-K1* to growth in medium containing 1% FBS.

Adaptation of *CHO-K1* to growth in medium with reduced serum concentration

- thaw cells and transfer them directly in prewarmed culture medium on day 0
- for the next weeks, passage cells according to table 10.1
- on day 47, freeze cells at $5 \cdot 10^6$ cells/ml in culture medium Ham-F12 containing 10% dimethyl sulfoxide (Sigma, D5879)

Pre-cultivation for experiments in reactor modules

CHO-K1 and *CHO-K1* adapted to growth in medium containing 1% FBS

- thaw cells and transfer them directly in prewarmed culture medium
- the next day, passage cells by a 1:5 split
- three days later, passage cells by a 1:4 split
- cells are used for inoculation the next day (exponential growth phase)

A549

- thaw cells and transfer them directly in prewarmed culture medium
- the next day, passage cells by a 1:5 split
- three days later, passage cells by a 1:4 split
- cells are used for inoculation the next day (exponential growth phase)

MDCK

- thaw cells and transfer them directly in prewarmed culture medium
- the next day, passage cells by a 1:5 split
- three days later, passage cells by a 1:4 split
- two days later, passage cells by a 1:3 split
- cells are used for inoculation two days later (exponential growth phase)

human primary fibroblasts

- thaw cells and transfer them directly in prewarmed culture medium
- three days later, passage cells by a 1:5 split
- about two weeks later, passage cells by a 1:4 split
- perform a medium replacement after one and four weeks of cultivation
- cells are used for inoculation six weeks after last passaging

Calculation of flow velocity

Superficial velocity v for flow through ceramics in reactor modules was calculated by taking into account porosity (ϕ) and perfused face area (A). Volumetric flow velocity $\dot{V}$ was calculated from pump characteristic and number of pump head revolutions.

$$v = \frac{\dot{V}}{A} \tag{10.1}$$

$$A = \pi * r_{foam}^2 * n_{foam} * \phi \tag{10.2}$$

n_{foam} and r_{foam} are number and radius of ceramic cylinders inside revolver magazine, respectively. For ceramics with drillings as in chapter 6.1.4 perfused face area was calculated as follows:

$$A = (A_{total} - A_{drillings}^0 + A_{drillings}) * \phi \tag{10.3}$$

$$A_{total} = \pi * r_{foam}^2 * n_{foam} \tag{10.4}$$

$$A_{drillings}^0 = \pi * r_{drilling}^2 * n_{drilling} * n_{foam} \tag{10.5}$$

$$A_{drillings} = 2\pi * r_{drilling} * h_{drilling} * n_{drilling} * n_{foam} + A_{drillings}^0 \tag{10.6}$$

$n_{drilling}$, $r_{drilling}$, and $h_{drilling}$ correspond to number of drillings per foam, radius, and height of each drilling, respectively.

Engineering drawings

The following pages contain engineering drawings of bioreactor devices used in this work, gently provided by A. Bischof, TU Berlin. These are in order of appearance:

- left shell of revolver module type A
- right shell of revolver module type A
- magazine type A
- revolver module type B
- left shell of revolver module type B
- right shell of revolver module type B

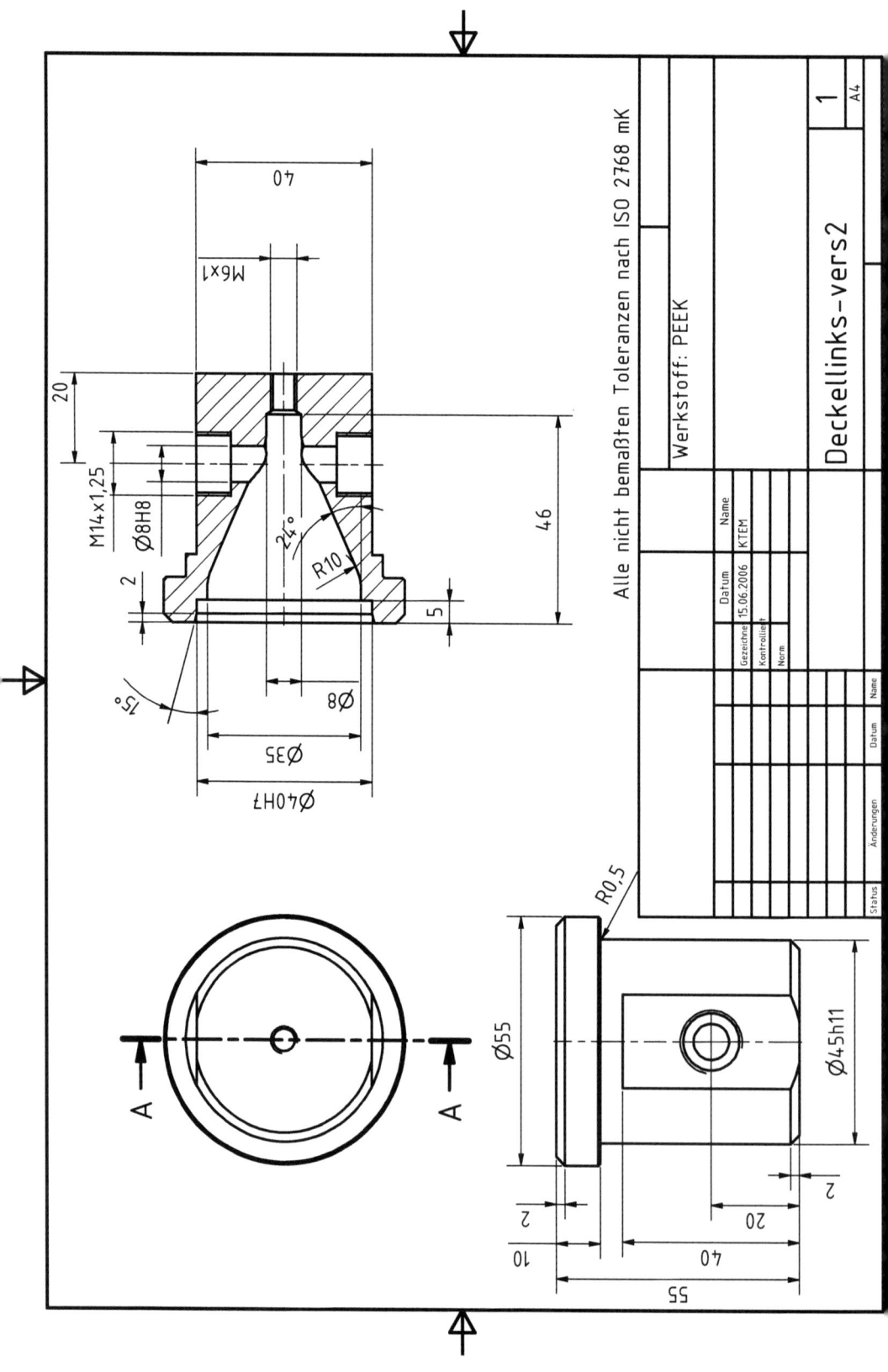

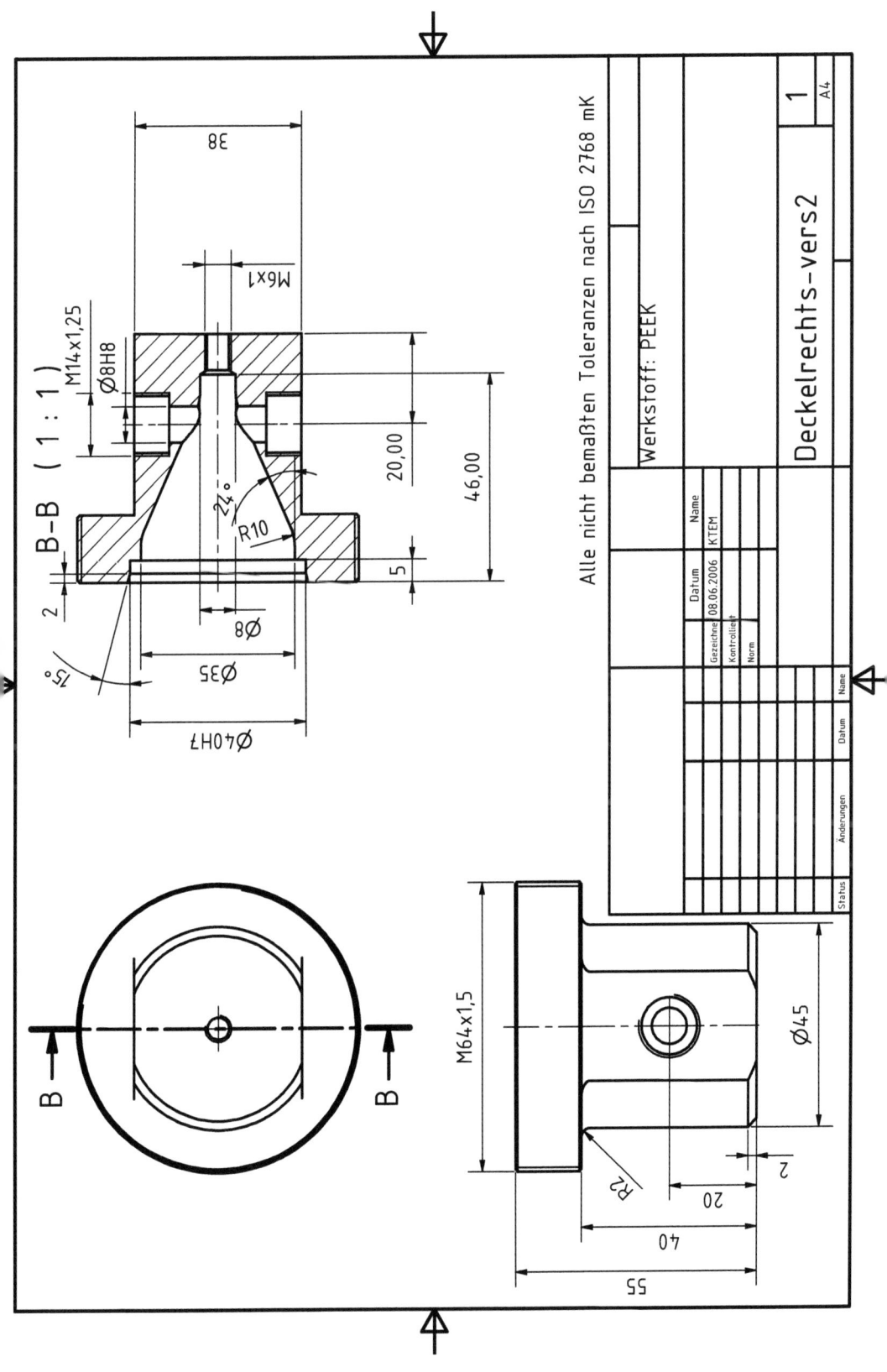

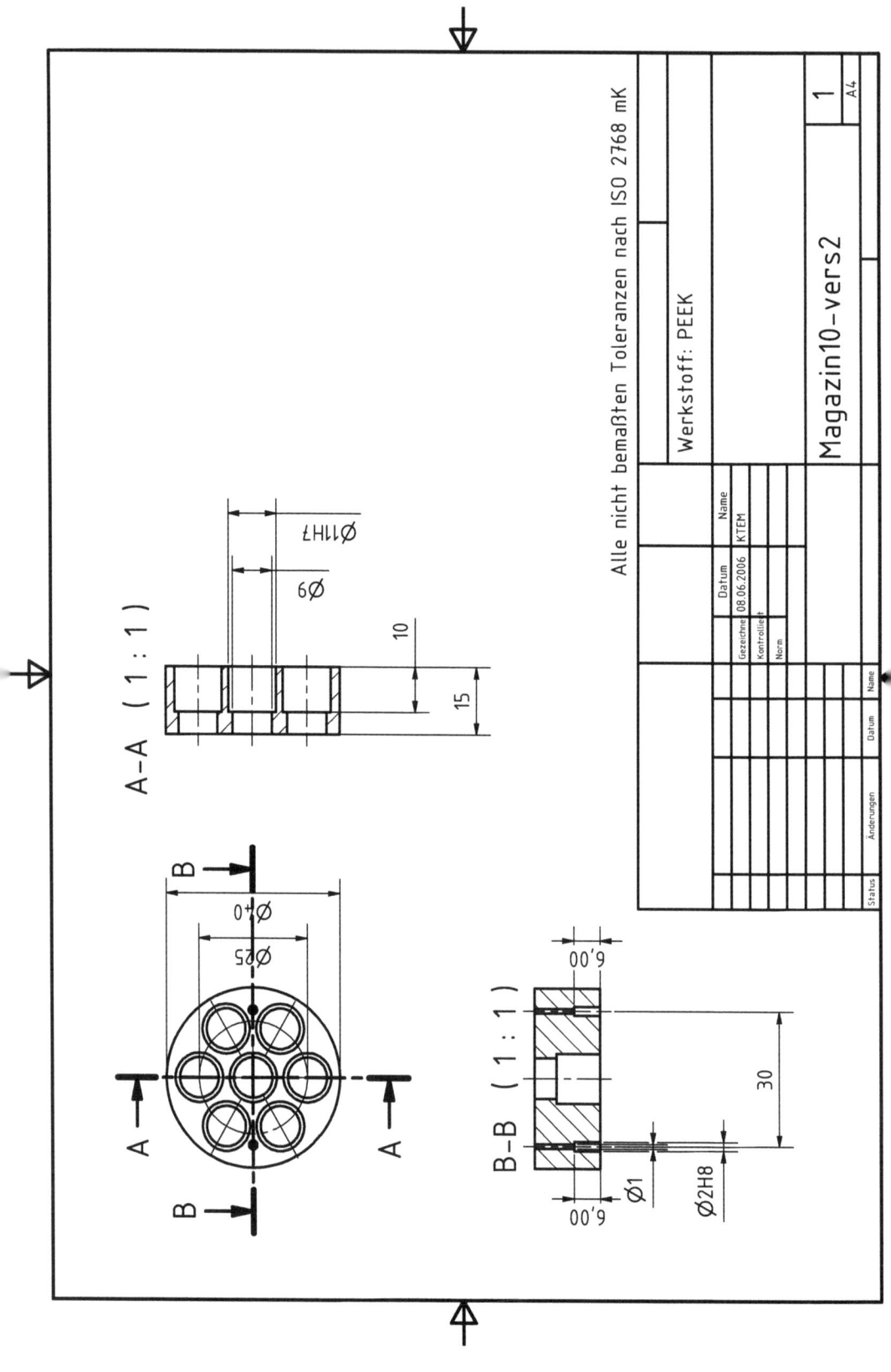

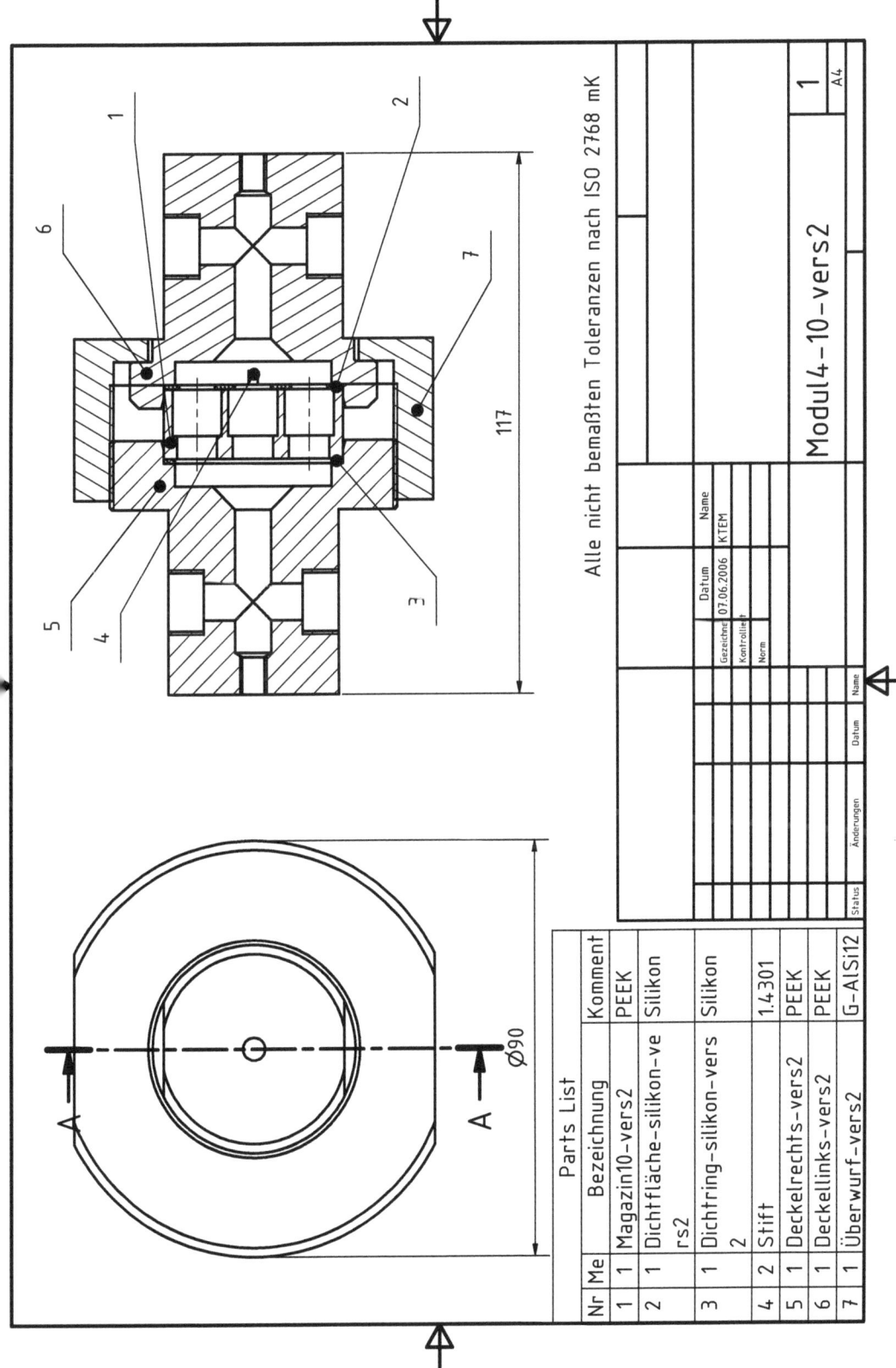

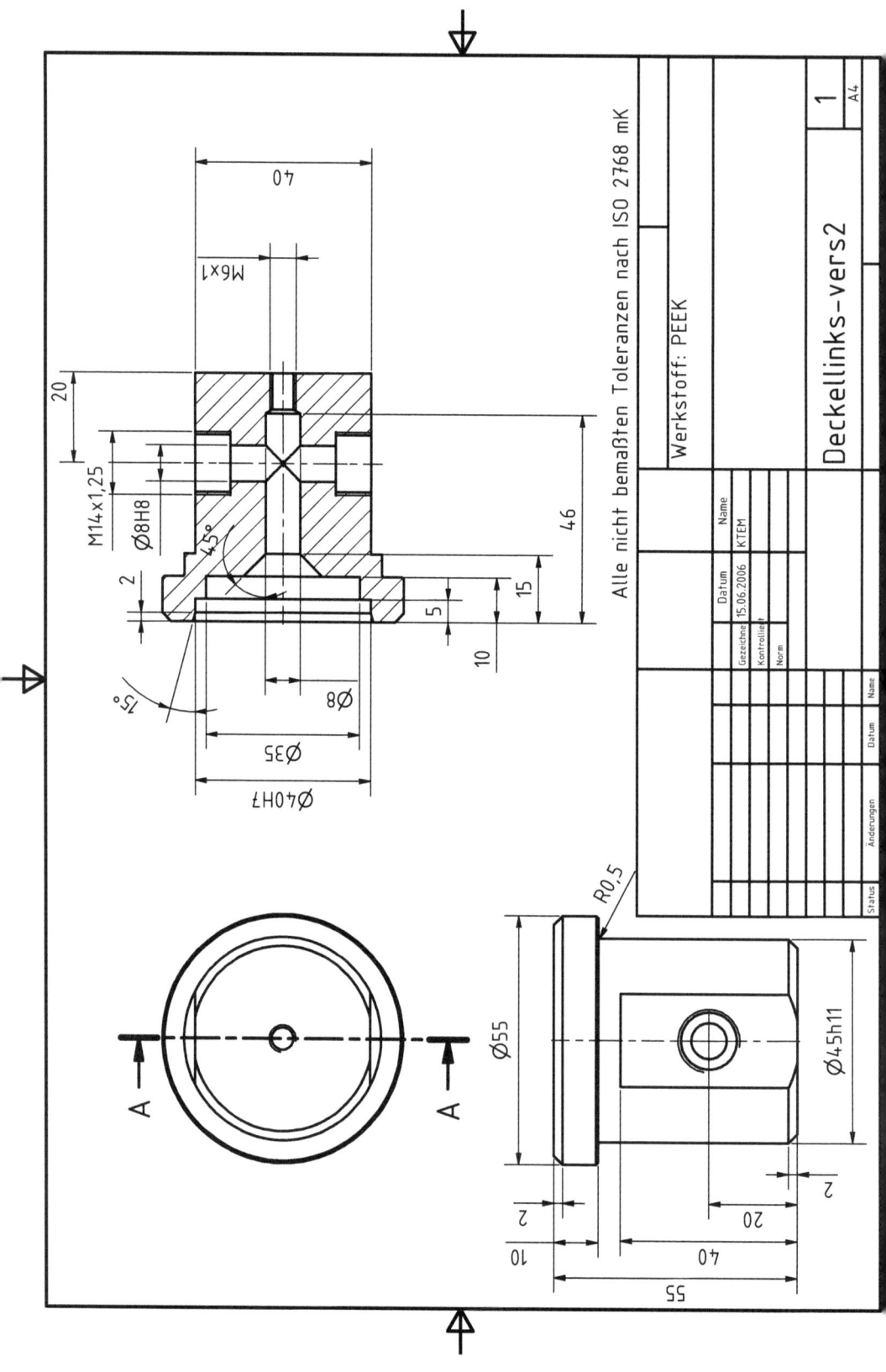

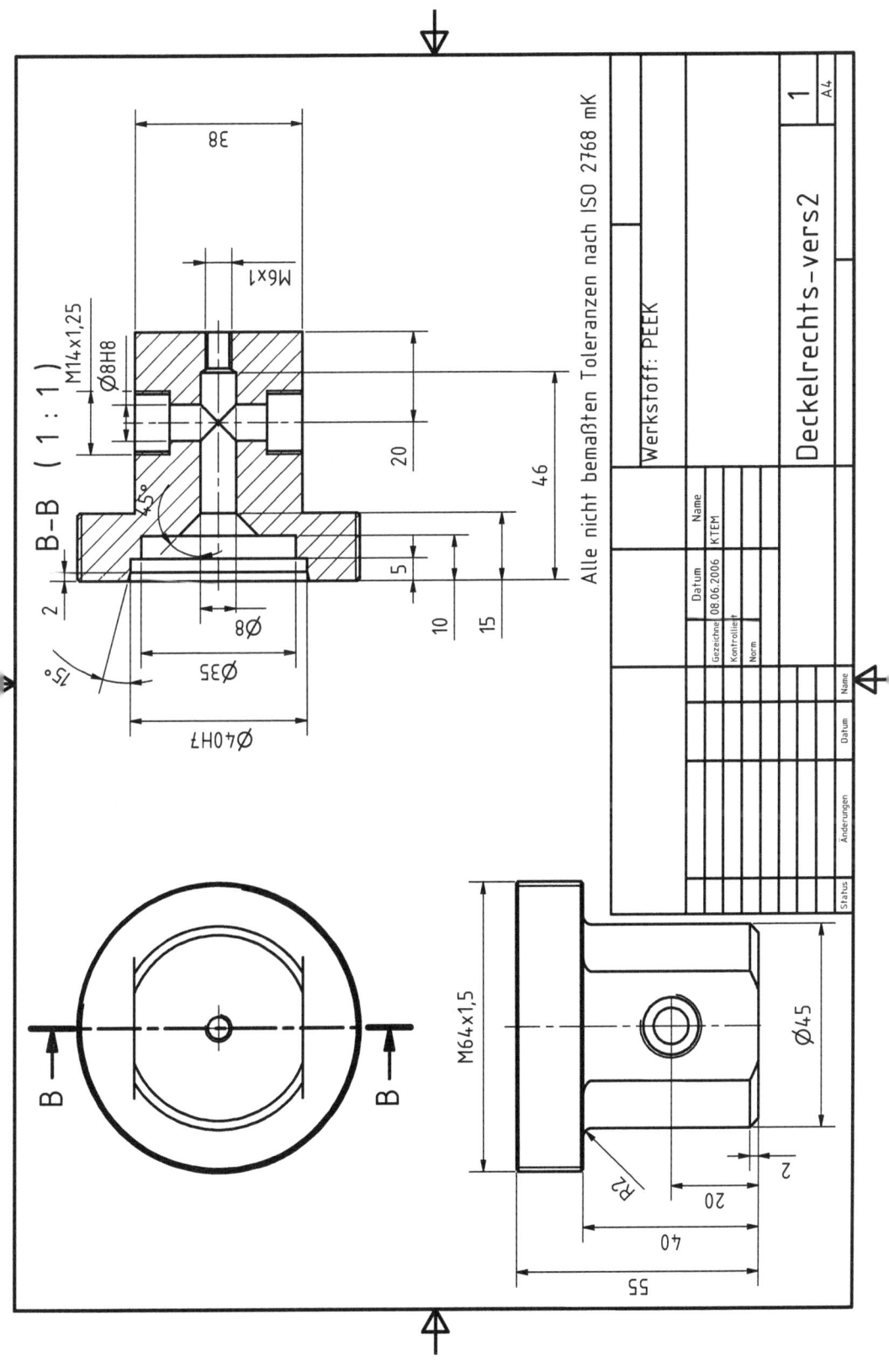

Literature

[Ahmed et al., 1994] Ahmed, S. A., Gogal, R. M., and Walsh, J. E. (1994). A new rapid and simple non-radioactive assay to monitor and determine the proliferation of lymphocytes: an alternative to [3H]thymidine incorporation assay. *J Immunol Methods*, 170(2):211–224.

[Alberts et al., 2007] Alberts, B., Johnson, A., Lewis, J., Raff, M., Roberts, K., and Walter, P. (2007). *Molecular Biology of the Cell*. Garland Science, Taylor & Francis Group, Fifth Edition.

[Altamirano et al., 2001] Altamirano, C., Illanes, A., Casablancas, A., Gamez, X., Cairo, J. J., and Godia, C. (2001). Analysis of CHO cells metabolic redistribution in a glutamate-based defined medium in continuous culture. *Biotechnol Prog*, 17:1032–1041.

[Anoopkumar-Dukie et al., 2005] Anoopkumar-Dukie, S., Carey, J. B., Conere, T., Sullivan, E. O., van Pelt, F. N., and Allshire, A. (2005). Resazurin assay of radiation response in cultured cells. *Br J Radiol*, 78:945–947.

[Anton et al., 2008] Anton, F., Suck, K., Diederichs, S., Behr, L., Hitzmann, B., van Griensven, M., Scheper, T., and Kasper, C. (2008). Design and characterization of a rotating bed system bioreactor for tissue engineering applications. *Biotechnol Prog*, 24:140–147.

[Applegate and Stephanopoulos, 1992] Applegate, M. and Stephanopoulos, G. (1992). Development of a single-pass ceramic matrix bioreactor for large-scale mammalian cell culture. *Biotechnol Bioeng*, 40:1056–1068.

[Bagley et al., 1999] Bagley, J., Rosenzweig, M., Marks, D. F., and Pykett, M. J. (1999). Extended culture of multipotent hematopoietic progenitors without cytokine augmentation in a novel three-dimensional device. *Exp Hematol*, 27 (3):496–504.

[Bancroft et al., 2002] Bancroft, G. N., Sikavitsast, V. I., van den Dolder, J., Sheffield, T. L., Ambrose, C. G., Jansen, J. A., and Mikos, A. G. (2002). Fluid flow increases mineralized matrix deposition in 3D perfusion culture of marrow stromal osteloblasts in a dose-dependent manner. *Proc Natl Acad Sci U S A*, 99(20):12600–12605.

[Bischof and Blessing, 2006] Bischof, A. and Blessing, L., editors (2006). *Geschäumte Keramikwerkstoffe: Neue Herausforderungen für den Produktentwicklungsprozess*, 17. Symposium "DESIGN FOR X", Neukirchen, 12.-13. Oktober 2006.

[Blan and Birla, 2008] Blan, N. R. and Birla, R. K. (2008). Design and fabrication of heart muscle using scaffold-based tissue engineering. *J Biomed Mater Res A*, 86(1):195–208.

[Bonassar and Vacanti, 1998] Bonassar, L. J. and Vacanti, C. A. (1998). Tissue engineering: The first decade and beyond. *J Cell Biochem Suppl*, 72(30-31):297–303.

[Burg et al., 2000] Burg, K. J. L., Jr., W. D. H., Culberson, C. R., Beiler, R. J., Greene, K. G., Loebsack, A. B., Roland, W. D., Eiselt, P., Mooney, D. J., and Halberstadt, C. R. (2000). Comparative study of seeding methods for three-dimensional polymeric scaffolds. *J Biomed Mater Res*, 51 (4):642–649.

[Carrier et al., 1999] Carrier, R. L., Papadaki, M., Rupnick, M., Schoen, F. J., Bursac, N., Langer, R., Freed, L. E., and Vunjak-Novakovic, G. (1999). Cardiac tissue engineering: Cell seeding, cultivation parameters, and tissue construct characterization. *Biotechnol Bioeng*, 64(5):580–589.

[Carrier et al., 2002] Carrier, R. L., Rupnick, M., Langer, R., Schoen, F. J., Freed, L. E., and Vunjak-Novakovic, G. (2002). Perfusion improves tissue architecture of engineered cardiac muscle. *Tissue Eng*, 8(2):175–188.

[Dar et al., 2002] Dar, A., Shachar, M., Leor, J., and Cohen, S. (2002). Optimization of cardiac cell seeding and distribution in 3D porous alginate scaffolds. *Biotechnol Bioeng*, 80(3):305–312.

[Davisson et al., 2002] Davisson, T., Sah, R. L., and Ratcliffe, A. (2002). Perfusion increases cell content and matrix synthesis in chondrocyte three-dimensional cultures. *Tissue Eng*, 8(5):807–816.

[Davoren et al., 2006] Davoren, M., Herzog, E., Casey, A., Cottineau, B., Chambers, G., Byrne, H. J., and Lyng, F. M. (2006). In vitro toxicity evaluation of single walled carbon nanotubes on human A549 lung cells. *Toxicol In Vitro*, 21(3):438–448.

[Deshpande and Heinzle, 2004] Deshpande, R. R. and Heinzle, E. (2004). On-line oxygen uptake rate and culture viability measurement of animal cell culture using microplates with integrated oxygen sensors. *Biotechnol Lett*, 26(9):763–767.

[Du et al., 2008] Du, D., Furukawa, K., and Ushida, T. (2008). Oscillatory perfusion seeding and culturing of osteoblast-like cells on porous beta-tricalcium phosphate scaffolds. *J Biomed Mater Res A*, 86A(3):796–803.

[Fassnacht and Poertner, 1999] Fassnacht, D. and Poertner, R. (1999). Experimental and theoretical considerations on oxygen supply for animal cell growth in fixed-bed reactors. *J Biotechnol*, 72:169–184.

[Freed and Vunjak-Novakovic, 1997] Freed, L. E. and Vunjak-Novakovic, G. (1997). Microgravity tissue engineering. *In Vitro Cell Dev Biol Anim*, 33(5):381–385.

[Garrn et al., 2004] Garrn, I., Reetz, C., Brandes, N., Kroh, L. W., and Schubert, H. (2004). Clot-forming: the use of proteins as binders for producing ceramic foams. *J Eur Ceram Soc*, 24(3):579–587.

[Genzel et al., 2004] Genzel, Y., Behrendt, I., Konig, S., Sann, H., and Reichl, U. (2004). Metabolism of MDCK cells during cell growth and influenza virus production in large-scale microcarrier culture. *Vaccine*, 22(17-18):2202–2208.

[Goldstein et al., 2001] Goldstein, A. S., Juarez, T. M., Helmke, C. D., Gustin, M. C., and Mikos, A. G. (2001). Effect of convection on osteoblastic cell growth and function in biodegradable polymer foam scaffolds. *Biomaterials*, 22(11):1279–1288.

[Gonzalez and Tarloff, 2001] Gonzalez, R. J. and Tarloff, J. B. (2001). Evaluation of hepatic subcellular fractions for Alamar blue and MTT reductase activity. *Toxicol In Vitro*, 15:257–259.

[Gooch et al., 2001] Gooch, K. J., Kwon, J. H., Blunk, T., Langer, R., Freed, L. E., and Vunjak-Novakovic, G. (2001). Effects of mixing intensity on tissue-engineered cartilage. *Biotechnol Bioeng*, 72(4):402–407.

[Goralczyk et al., 2009] Goralczyk, V., Driemel, G., Kroh, L. W., and King, R. (2009). Zellzahlbestimmung in dreidimensionalen, porösen Gerüststrukturen. *CIT Special Issue: Process-Net-Jahrestagung und 27. Jahrestagung der Biotechnologen*, 81(8):1271.

[Grampp et al., 1996] Grampp, G. E., Applegate, M. A., and Stephanopoulos, G. (1996). Cyclic operation of ceramic-matrix animal cell bioreactors for controlled secretion of an endocrine hormone. a comparison of single-pass and recycle modes of operation. *Biotechnol Prog*, 12:837–846.

[Habisreuther et al., 2008] Habisreuther, P., Djordjevic, N., and Zarzalis, N. (2008). Numerische Simulation der Mikroströmung in porösen inerten Strukturen. *Chemie Ingenieur Technik*, 80(3):327–341.

[Hagen, 2005] Hagen, J. (2005). *Chemiereaktoren; Nichtideale Reaktoren und Reaktormodelle*, chapter 7. Wiley.

[Hansen and Emborg, 1994] Hansen, H. A. and Emborg, C. (1994). Influence of ammonium on growth, metabolism, and productivity of a continuous suspension chinese hamster ovary cell culture. *Biotechnol Prog*, 10:121–124.

[Harada et al., 1996] Harada, Y., Wang, J. T., Doppalapudi, V. A., Willis, A. A., Jasty, M., Harris, W. H., Nagase, M., and Goldring, S. R. (1996). Differential effects of different forms of hydroxyapatite and hydroxyapatite tricalcium phosphate particulates on human monocyte macrophages in vitro. *J Biomed Mater Res*, 31(1):19–26.

[Holtorf et al., 2005] Holtorf, H. L., Sheffield, T. L., Ambrose, C. G., Jansen, J. A., and Mikos, A. G. (2005). Flow perfusion culture of marrow stromal cells seeded on porous biphasic calcium phosphate ceramics. *Ann Biomed Eng*, 33(9):1238–1248.

[Holy et al., 2000] Holy, C., Shoichet, M. S., and Davies, J. E. (2000). Engineering three-dimensional bone tissue in vitro using biodegradable scaffolds: Investigating initial cell-seeding density and culture period. *J Biomed Mater Res*, 51 (3):376–382.

[Innocentini et al., 1998] Innocentini, M. D. M., Salvini, V. R., Pandolfelli, V. C., and Coury, J. R. (1998). The permeability of ceramic foams. *www.ceramicbulletin.org*, 78 (9).

[Janssen et al., 2006] Janssen, F. W., Hofland, I., van Oorschot, A., Oostra, J., Peters, H., and van Blitterswijk, C. A. (2006). Online measurement of oxygen consumption by goat bone marrow stromal cells in a combined cell-seeding and proliferation perfusion bioreactor. *J Biomed Mater Res A*, 79(2):338–348.

[Kim et al., 1997] Kim, B.-S., Putnam, A. J., Kulik, T. J., and Mooney, D. J. (1997). Optimizing seeding and culture methods to engineer smooth muscle tissue on biodegradable polymer matrices. *Biotechnol Bioeng*, 57(1):46–54.

[Kitagawa et al., 2006] Kitagawa, T., Yamaoka, T., Iwase, R., and Murakami, A. (2006). Three-dimensional cell seeding and growth in radial-flow perfusion bioreactor for in vitro tissue reconstruction. *Biotechnol Bioeng*, 93(5):947–954.

[Knazek et al., 1972] Knazek, R., Gullino, P., Kohler, P., and Dedrick, R. (1972). Cell culture on artificial capillaries: an approach to tissue growth in vitro. *Science*, 178(56):65–66.

[Krajewski et al., 1996] Krajewski, A., Ravaglioli, A., Kirsch, M., Biagini, G., Solmi, R., Belmonte, M., Zucchini, C., Gandolfi, M. G., Castaldini, C., Rodriguez, L., Giardino, R., Mongiorgi, R., Roncari, E., and Orlandi, L. (1996). Ceramic support for cell cultures. *J Mater Sci Mater Med*, 7:99–102.

[Ku et al., 1981] Ku, K., Kuo, M. J., Delente, J., Wildi, B. S., and Feder, J. (1981). Development of a hollow-fiber system for large-scale culture of mammalian-cells. *Biotechnol Bioeng*, 23(1):79–95.

[Leach and Schmidt, 2004] Leach, J. B. and Schmidt, C. E., editors (2004). *Photocrosslinkable Hyaluronic Acid Hydrogels for Tissue Engineering*, Mat Res Soc Symp Proc EXS-1.

[Lee et al., 1991] Lee, D. W., Grace, J. R., Chow, B. K. C., MacGillivray, R. T. A., and Kilburn, D. G. (1991). High density cultivation of BSK cells on sintered alumina ceramic foam support. *Cytotechnology*, 5:233–241.

[LeGeros, 2002] LeGeros, R. Z. (2002). Properties of osteoconductive biomaterials: Calcium phosphates. *Clin Orthop Relat Res*, 395:81–98.

[Li et al., 2001] Li, Y., Ma, T., Kniss, D. A., Lasky, L. C., and Yang, S.-T. (2001). Effects of filtration seeding on cell density, spatial distribution, and proliferation in nonwoven fibrous matrices. *Biotechnol Prog*, 17(5):935–944.

[Li et al., 2008] Li, Z., Gunn, J., Chen, M.-H., Cooper, A., and Zhang, M. (2008). On-site alginate gelation for enhanced cell proliferation and uniform distribution in porous scaffolds. *J Biomed Mater Res A*, 86(2):552–559.

[Liao and Cui, 2004] Liao, S. S. and Cui, F. Z. (2004). In vitro and in vivo degradation of mineralized collagen-based composite scaffold: nanohydroxyapatite/collagen/poly(L-lactide). *Tissue Eng*, 10(1-2):73–80.

[Lieber et al., 1976] Lieber, M., Smith, B., Szakal, A., Nelson-Rees, W., and Todaro, G. (1976). A continuous tumor-cell line from a human lung carcinoma with properties of type II alveolar epithelial cells. *Int J Cancer*, 17:62–70.

[Lydersen, 1987] Lydersen, B. K. (1987). *Perfusion Cell Culture System Based on Ceramic Matrices, in "Large Scale Cell Culture Technology"*. B. K. Lydersen.

[Lydersen et al., 1985] Lydersen, B. K., Pugh, G. G., Paris, M. S., Sharma, B. P., and Noll, L. A. (1985). Ceramic matrix for large scale animal cell culture. *Nat Biotechnol*, 3:63–67.

[Marler et al., 1998] Marler, J. J., Upton, J., Langer, R., and Vacanti, J. P. (1998). Transplantation of cells in matrices for tissue regeneration. *Adv Drug Deliv Rev*, 33:165–182.

[Martin and Vermette, 2005] Martin, Y. and Vermette, P. (2005). Bioreactors for tissue mass culture: design, characterization, and recent advances. *Biomaterials*, 26(35):7481–7503.

[Mehta and Linderman, 2005] Mehta, K. and Linderman, J. J. (2005). Model-based analysis and design of a microchannel reactor for tissue engineering. *Biotechnol Bioeng*, 94(3):596–609.

[Meuwly et al., 2007] Meuwly, F., Ruffieux, P.-A., Kadouri, A., and von Stockar, U. (2007). Packed-bed bioreactors for mammalian cell culture: Bioprocess and biomedical applications. *Biotechnol Adv*, 25(1):45–56.

[Möhler et al., 2008] Möhler, L., Bock, A., and Reichl, A. (2008). Segregated mathematical model for growth of anchorage-dependent MDCK cells in microcarrier culture. *Biotechnol Prog*, 24:110–119.

[Minuth et al., 2003] Minuth, W. W., Strehl, R., and Schumacher, K. (2003). *Zukunftstechnologie Tissue Engineering*. Wiley-VCH.

[Mitsuda et al., 1991] Mitsuda, S., Matsuda, Y., Kobayashi, N., Suzuki, A., Itagaki, Y., Kumazawe, E., Higashio, K., and Kawanishi, G. (1991). Continuous production of tissue plasminogen activator (t-PA) by human embryonic lung diploid fibroblast, IMR-90 cells, using a ceramic bed reactor. *Cytotechnology*, 6(1):23–31.

[Moutos et al., 2007] Moutos, F. T., Freed, L. E., and Guilak, F. (2007). A biomimetic three-dimensional woven composite scaffold for functional tissue engineering of cartilage. *Nat Mater*, 6(2):162–167.

[Narula et al., 1998] Narula, P., Xu, J., Kazzaz, J. A., Robbins, C. G., Davis, J. M., and Horowitz, S. (1998). Synergistic cytotoxicity from nitric oxide and hyperoxia in cultured lung cells. *Am J Physiol Lung Cell Mol Physiol*, 274(3):L411–L416.

[Navarro et al., 2001] Navarro, F. A., Mizuno, S., Huertas, J. C., Glowacki, J., and Orgill, D. P. (2001). Perfusion of medium improves growth of human oral neomucosal tissue constructs. *Wound Repair Regen*, 9(6):507–512.

[Neves et al., 2005] Neves, A. A., Medcalf, N., and Brindle, K. M. (2005). Influence of stirring-induced mixing on cell proliferation and extracellular matrix deposition in meniscal cartilage constructs based on polyethylene terephthalate scaffolds. *Biomaterials*, 26:4828–4836.

[O'Brien et al., 2000] O'Brien, J., Wilson, I., Orton, T., and Pognan, F. (2000). Investigation of the alamar blue (resazurin) fluorescent dye for the assessment of mammalian cell cytotoxicity. *Eur J Biochem*, 276:5421–5426.

[Orlandi et al., 1997] Orlandi, L., Solmi, R., Krajewski, A., Bearzatto, A., Biagini, G., Ciccopiedi, E., and Ravaglioli, A. (1997). Cell growth on cordierite: an approach to the identification of reliable supports for continuous-flow solid-bed reactors. *Biomaterials*, 18(14):955–961.

[Ouyang and Yang, 2007] Ouyang, A. and Yang, S.-T. (2007). Effects of mixing intensity on cell seeding and proliferation in three-dimensional fibrous matrices. *Biotechnol Bioeng*, 96(2):371–380.

[Park and Stephanopoulos, 1993] Park, S. J. and Stephanopoulos, G. (1993). Packed-bed bioreactor with porous ceramic beads for animal-cell culture. *Biotechnol Bioeng*, 41(1):25–34.

[Pazzano et al., 2000] Pazzano, D., Mercier, K. A., Moran, J. M., Fong, S. S., DiBiasio, D. D., Rulfs, J. X., Kohles, S. S., and Bonassar, L. J. (2000). Comparison of chondrogensis in static and perfused bioreactor culture. *Biotechnol Prog*, 16(5):893–896.

[Piret et al., 1991] Piret, J. M., Devens, D. A., and Cooney, C. L. (1991). Nutrient and metabolite gradients in mammalian cell hollow fiber bioreactors. *Can J Chem Eng*, 69:421–428.

[Radisic et al., 2003] Radisic, M., Euloth, M., Yang, L., Langer, R., Freed, L. E., and Vunjak-Novakovic, G. (2003). High-density seeding of myocyte cells for cardiac tissue engineering. *Biotechnol Bioeng*, 82(4):403–414.

[Schubert et al., 2004] Schubert, H., Garrn, I., Berthold, A., Knauf, W. U., Reufi, B., Fietz, T., and Gross, U. M. (2004). Culture of haematopoietic cells in a 3-D bioreactor made of Al2O3 or apatite foam. *J Mater Sci Mater Med*, 15(4):331–334.

[Scudiero et al., 1988] Scudiero, D. A., Shoemaker, R. H., Paull, K. D., Monks, A., Tierney, S., Nofziger, T. H., Currens, M. J., Seniff, D., and Boyd, M. R. (1988). Evaluation of a soluble tetrazolium formazan assay for cell-growth and drug sensitivity in culture using human and other tumor-cell lines. *Cancer Res*, 48(17):4827–4833.

[Sodian et al., 2002] Sodian, R., Lemke, T., Fritsche, C., Hoerstrup, S. P., Fu, P., Potapov, E. V., Hausmann, H., and Hetzer, R. (2002). Tissue-engineering bioreactors: A new combined cell-seeding and perfusion system for vascular tissue engineering. *Tissue Engineering*, 8(5):863–870.

[Suck et al., 2008] Suck, K., Fischer, M., van Griensven, M., Stahl, F., Scheper, T., Kasper, C., Hoffmeister, H., and Behr, L. (2008). Cultivation of MC3T3-E1 cells on a newly developed material (Sponceram®) using a rotating bed system bioreactor. *J Biomed Mater Res A*, 80A(2):268–275.

[Sutherland et al., 1986] Sutherland, R. M., Sordat, B., Bamat, J., Gabbert, H., Bourrat, B., and Mueller-Klieser, W. (1986). Oxygenation and differentiation in multicellular spheroids of human colon carcinoma. *Cancer Res*, 46:5320–5329.

[Suzuki et al., 1994] Suzuki, T., Sato, T., and Kominami, M. (1994). A dense cell retention culture system using a stirred ceramic membrane reactor. *Biotechnol Bioeng*, 44:1186–1192.

[Tharakan and Chau, 1986] Tharakan, J. P. and Chau, P. C. (1986). A radial flow hollow fiber bioreactor for the large-scale culture of mammalian-cells. *Biotechnol Bioeng*, 28(3):329–342.

[Timmins et al., 2007] Timmins, N. E., Scherberich, A., Frueh, J.-A., Heberer, M., Martin, I., and Jakob, M. (2007). Three-dimensional cell culture and tissue engineering in a T-CUP (Tissue Culture Under Perfusion). *Tissue Eng*, 13(8):2021–2028.

[Voytik-Harbin et al., 1998] Voytik-Harbin, S. L., Brightman, A. O., Waisner, B., and Lamar, C. H. (1998). Application and evaluation of the alamarblue assay for cell growth and survival of fibroblasts. *In Vitro Cell Dev Biol Anim*, 34:239–246.

[Vunjak-Novakovic et al., 1996] Vunjak-Novakovic, G., Freed, L. E., Biron, R. J., and Langer, R. (1996). Effects of mixing on the composition and morphology of tissue-engineered cartilage. *AIChE J*, 42(3):850–860.

[Vunjak-Novakovic et al., 1998] Vunjak-Novakovic, G., Obradovic, B., Martin, I., Bursac, P. M., Langer, R., and Freed, L. E. (1998). Dynamic cell seeding of polymer scaffolds for cartilage tissue engineering. *Biotechnol Prog*, 14:193–202.

[Wang et al., 2003] Wang, Y. C., Uemura, T., Dong, R., H., T., Kojima, J., and Tateishi, T. (2003). Application of perfusion culture system improves in vitro and in vivo osteogenesis of bone marrow-derived osteoblastic cells in porous ceramic materials. *Tissue Eng*, 9(6):1205–1214.

[Wendt et al., 2003] Wendt, D., Marsano, A., Jakob, M., Heberer, M., and Martin, I. (2003). Oscillating perfusion of cell suspensions through three-dimensional scaffolds enhances cell seeding efficiency and uniformity. *Biotechnol Bioeng*, 84 (2):205–214.

[Wendt et al., 2006] Wendt, D., Stroebel, S., Jakob, M., John, G. T., and Martin, I. (2006). Uniform tissues engineered by seeding and culturing cells in 3d scaffolds under perfusion at defined oxygen tensions. *Biorheology*, 43(3):481–488.

[Wickramasinghe, 1975] Wickramasinghe, S. N. (1975). *Human Bone Marrow*. Blackwell Scientific, Oxford.

[Wu et al., 1992] Wu, P., Ray, N. G., and Shuler, M. L. (1992). A single-cell model for CHO cells. *Ann NY Acad Sci*, 665:152–187.

[Xiao et al., 1999] Xiao, Y.-L., Riesle, J., and van Blitterswijk, C. A. (1999). Static and dynamic fibroblast seeding and cultivation in porous PEO/PBT scaffolds. *J Mater Sci Mater Med*, 10:773–777.

[Zhao and Ma, 2005] Zhao, F. and Ma, T. (2005). Perfusion bioreactor system for human mesenchymal stem cell tissue engineering: Dynamic cell seeding and construct development. *Biotechnol Bioeng*, 91 (4):482–493.

I want morebooks!

Buy your books fast and straightforward online - at one of world's fastest growing online book stores! Environmentally sound due to Print-on-Demand technologies.

Buy your books online at
www.morebooks.shop

Kaufen Sie Ihre Bücher schnell und unkompliziert online – auf einer der am schnellsten wachsenden Buchhandelsplattformen weltweit! Dank Print-On-Demand umwelt- und ressourcenschonend produziert.

Bücher schneller online kaufen
www.morebooks.shop

KS OmniScriptum Publishing
Brivibas gatve 197
LV-1039 Riga, Latvia
Telefax:+371 686 204 55

info@omniscriptum.com
www.omniscriptum.com

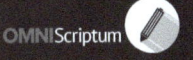

Printed by Books on Demand GmbH, Norderstedt / Germany